WITHDRAWN

Carnegie Mellon

Benefit-Cost Analysis of Air-Pollution Control

Benefit-Cost Analysis of Air-Pollution Control

Robert Halvorsen
Michael G. Ruby
University of Washington

LexingtonBooks
D.C. Heath and Company
Lexington, Massachusetts
Toronto

363.7392
H19b

Library of Congress Cataloging in Publication Data

Halvorsen, Robert.
  Benefit-cost analysis of air-pollution control.

  Includes bibliographical references and index.
  1. Air quality management—Cost effectiveness. I. Ruby, Michael G. II. Title.
TD883.H35   363.7'3925   78-19587
ISBN 0-669-02647-6   AACR2

Copyright © 1981 by D.C. Heath and Company

All rights reserved. No part of this publication may be reproduced or transmitted in any form or by any means, electronic or mechanical, including photocopy, recording, or any information storage or retrieval system, without permission in writing from the publisher.

Published simultaneously in Canada

Printed in the United States of America

International Standard Book Number: 0-669-02647-6

Library of Congress Catalog Card Number: 78-19587

For S.A.T. and E.L.R.

Contents

	List of Figures	xi
	List of Tables	xiii
	Preface and Acknowledgments	xv
Chapter 1	**Introduction**	1
	Optimal Degree of Air-Pollution Control	2
	Benefit-Cost Analysis	5
	Criticisms of Benefit-Cost Analysis	8
	Plan of the Book	9
Part I	*The Theory of Benefit-Cost Analysis*	11
Chapter 2	**Theoretical Foundations of Benefit-Cost Analysis**	13
	Social Welfare	13
	Economic-Efficiency Criterion	14
	Social-Welfare Criterion	17
	Economic-Welfare Criterion	19
	Potential-Compensation Criteria	20
	Comparison of Potential-Compensation and Economic-Welfare Criteria	21
	Potential Economic Welfare	23
	Summary	25
Chapter 3	**Aggregation of Benefits and Costs over Individuals**	29
	Aggregation without Explicit Distributional Weights	30
	Aggregation with Explicit Distributional Weights	32
	Summary	35
Chapter 4	**Aggregation of Benefits and Costs over Time**	37
	Optimization over Time	37
	The Concept of Dynamic Efficiency	38
	Present-Value Criterion	41

	Net-Present-Value Decision Rules	45
	Alternative Investment-Decision Rules	46
	Comparison of Decision Rules	52
	Sources of Dynamic Inefficiency	52
	Implications of Dynamic Inefficiency	54
	Net-Social-Benefit Criterion	57
	Summary	60
Chapter 5	**Evaluation of Uncertain Benefits and Costs**	63
	Expected Monetary Value	63
	Expected Utility	65
	Prospect Theory	69
	Implications for Policy Evaluation	70
	Summary	74
Chapter 6	**Valuation of Priced and Unpriced Commodities**	77
	Perfectly Competitive Markets	77
	Imperfectly Competitive Markets	83
	Nonexistent Markets	85
	Summary	94
Part II	*Application to Air-Pollution Control*	97
Chapter 7	**Quantifying Air-Pollution Effects**	99
	Estimating Pollutant Exposure	99
	Dose-Response Functions	104
	Multivariate Regression Analysis	107
	Sources of Error in Regression Analysis	108
	Summary	110
Chapter 8	**Estimating Health Benefits**	111
	Development of Health-Effects Dose-Response Functions	111
	Macroepidemiologic Studies of Mortality and Morbidity	113
	Subjective Assessment of Dose-Response Functions	122
	Valuation of Health Damages	124
	Summary	129
Chapter 9	**Estimating Vegetation and Ecosystems Benefits**	133
	Vegetation Dose-Response Functions	133

Contents ix

	Animal and Ecosystems Dose-Response Functions	140
	Valuation of Vegetation, Animal, and Ecosystem Damages	141
	Summary	142
Chapter 10	**Estimating Materials Benefits**	**145**
	Metals Dose-Response Functions	146
	Paint Dose-Response Functions	149
	Textiles Dose-Response Functions	150
	Other Materials Damages	151
	Valuation of Materials Damages	152
	Summary	153
Chapter 11	**Estimating Aesthetic Benefits**	**155**
	Odor Damages	155
	Soiling Damages	155
	Valuation of Soiling Damages	157
	Visibility Damages	159
	Valuation of Visibility Damages	161
	Property-Value Studies	163
	Summary	166
Chapter 12	**Procedures for Evaluation of Control Costs**	**169**
	Types of Cost Estimates	169
	Defining Control Requirements	171
	Items Included in Cost Estimates	172
	Sources of Variability in Estimates	174
	Price Adjustments to Cost Estimates	176
	Financial Analysis of Control Costs	178
	Summary	179
Chapter 13	**Data for Estimating Control Costs**	**181**
	Cost Functions for Order-of-Magnitude Estimates	181
	Making a Study Estimate	189
	Other Cost Data	191
	Estimating Transportation-Pollutants-Control Costs	194
Appendix A	**Sources, Measurement, and Effects of Air Pollutants**	**197**

Appendix B	**Air-Pollution-Control Equipment**	215
Appendix C	**Order-of-Magnitude Cost Functions**	229
	References	237
	Index	257
	About the Authors	265

List of Figures

1-1	Optimal Degree of Air-Pollution Control	3
1-2	Steps in Benefit-Cost Analysis	6
2-1	Utility-Possibility Frontier	15
2-2	Maximization of Social Welfare	18
3-1	Distributional Preferences and Ranking of Policies	34
4-1	Maximization of Utility over Time	39
4-2	Discount Rates and Ranking of Investments	51
5-1	Utility Function for Decisions Involving Uncertainty	67
5-2	Approximation Errors Using Expected-Monetary-Value Decision Rules	72
6-1	Individual's Demand Curve for a Commodity	78
6-2	Individual's Labor-Supply Curve	81
6-3	Market Demand and Supply Curves for a Commodity	82
6-4	Valuation of Large Changes	83
6-5	Effect of Improvement in Environmental Quality	86
6-6	Alternative Supply Assumptions	92
7-1	Gaussian Plume Model	101
7-2	Dose-Response Functions	106
8-1	Subjective Dose-Response Curves for Excess Angina Attacks from Exposure to Carbon Monoxide	125
9-1	Reductions in Yield for Four Crops	136
11-1	Visibility as a Function of Fine-Particle Concentrations	162
B-1	Cyclone Mechanical Collector for Particulate Matter	217
B-2	Electrostatic Precipitator for Particulate Matter	219
B-3	Fabric Filters for Particulate Matter	221
B-4	Venturi Scrubber for Particulate Matter	223
B-5	Packed-Tower Absorber for Gases	225

List of Tables

1-1	Estimated Annual Benefits and Costs of Air-Pollution Control in the United States	2
2-1	Illustration of Potential-Compensation Criterion	24
3-1	Effect on Economic Welfare of Policy A	32
4-1	Calculation of Net Present Value	46
4-2	Alternative Investment-Decision Rules	48
4-3	Internal Rates of Return	50
4-4	Calculation of the Social Value of Private Investment	56
4-5	Ratio of V_C to V_B When $V = \$6.00$	60
5-1	Aggregation over States of the World	63
6-1	Multiplicative Factors for Estimating Total Consumers' and Producers' Surplus	93
8-1	Macroepidemiologic Studies of Air Pollution and Mortality in the United States	114
8-2	Nelson, Knelson, and Hasselblad Dose-Response Functions	123
8-3	Estimated Expenditures on Illness	127
9-1	Estimated Coefficients for Heck Ozone Dose-Response Functions	137
10-1	Relative Economic Importance of Pollutants in Materials Damage	146
10-2	Steel Products Subject to Pollution Damage in 1970	148
11-1	Estimated Coefficients for Beloin and Haynie Soiling Dose-Response Functions	157
11-2	Household-Cleaning-Operation Unit Costs	159
11-3	Mean Willingness to Pay for Visibility	164
12-1	Expenditures Included in Cost Analysis	173
12-2	Price Indexes for Air-Pollution-Control Projects	177

13-1	Scaling Factors for the Purchase Cost of Air-Pollution-Control Equipment	182
13-2	Air-Pollution-Control Equipment-Cost Functions	185
13-3	Coefficients for Burchard–Marder Cost Functions	187
13-4	Air-Pollution-Monitoring Network and Intermittent-Control-System Costs	193
B-1	Average Annual U.S. Purchases of Air-Pollution-Control Equipment by Industry	216
C-1	Cost of Clean Air and SEAS Cost Functions	230

Preface and Acknowledgments

The scope and impact of air-pollution-control policies have increased dramatically during the last two decades. In the United States, the federal government obtained enforcement powers with respect to air-pollution control only in 1963 with the passage of the Clean Air Act, and its present extensive role in setting and enforcing air-quality standards dates mainly from the 1970 Clean Air Amendments. Despite the short history of legislated air-pollution control in the United States, the impact on air quality has been substantial. During this period the U.S. urban population has increased by 40 percent, and industrial output has doubled; yet, with a few notable exceptions, urban air-pollution concentrations have declined substantially. However, the increased costs of air-pollution control have been a source of concern, especially with regard to the possible effects of air-pollution control on the goal of energy independence.

This book discusses the use of benefit-cost analysis to evaluate alternative air-pollution-control policies. The goal of the book is to make clear the principles underlying practical benefit-cost analysis. Because the intended audience for the book includes both economists and air-pollution-control specialists, we have attempted to minimize the use of each profession's specialized vocabulary and to define unfamiliar terms when they do appear. As a result, some explanations may seem overly long to readers already familiar with the topic being discussed. We hope this cost is justified by the benefit of greater understanding by those less familiar with the material.

We are grateful to a number of people for their generous assistance in the preparation of this book. Reviewers of earlier drafts of portions of the manuscript include A. Basala, D. Gillette, V. Hasselblad, F. Haynie, D. McKee, and W. Riggan of the U.S. Environmental Protection Agency; A. Cohen of CIS Resource Systems; B. Felske of the Economic Council of Canada; W. Heck of North Carolina State University; R. Mendelsohn and R. Zerbe of the University of Washington; J. Thielke of Puget Sound Power and Light; V. Salvin of the University of North Carolina, Greensboro; V. Uhl of the University of Virginia; and D. Vichoreck of the Montana Air Quality Bureau. We retain responsibility for any remaining errors.

The too numerous drafts of the manuscript were typed by Marian Bolan. As always, her speed, accuracy, and patience were remarkable. We are also grateful to Dale Leuthold for the skill with which she prepared the graphs and illustrations.

Benefit-Cost Analysis of Air-Pollution Control

1 Introduction

The effects of an air pollutant may include increased illness and earlier death, deterioration of materials, reduced agricultural productivity, reduced visibility, increased soiling, and annoying odors. Air-pollution control can benefit society by reducing these damages, but control programs require resources that could be used to produce other goods and services. The choice of an appropriate air-pollution-control policy requires comparison of the benefits and the costs of alternative policies in order to determine the net impact of each policy on social well-being.

The benefits and costs of an air-pollution-control policy may take many different forms, occur at different times, involve different degrees of uncertainty, and affect different individuals. The economic theory of benefit-cost analysis provides a framework for taking such diversity into account. This book discusses the theoretical and practical issues involved in using benefit-cost analysis to evaluate air-pollution-control policies.

The importance of making the correct policy choices is suggested by the magnitude of the benefits and costs of air-pollution control. Recent estimates of the total annual benefits and costs of air-pollution control in the United States are shown in table 1-1. The figures for benefits are Freeman's (1979a) estimates of the reduction in each type of damages between 1970 and 1978. Freeman emphasizes the uncertainty underlying these estimates and suggests that, although the most reasonable point estimate of total benefits is $21.4 billion, the actual value may be as low as $4.6 billion or as high as $51.2 billion. The estimate of $16.7 billion for annual costs is calculated from survey and census data for 1977 (Rutledge 1979; Rutledge and O'Connor 1979).

Because the estimates of benefits and costs in table 1-1 do not distinguish between geographic areas, are neither complete nor fully comparable, and do not indicate the benefits and costs of alternative degrees of control, they are not sufficient to evaluate the desirability of existing air-pollution-control policies.[1] The determination of the optimal degree of control, and the question of whether government action is required to attain it, are discussed in the following section.

1

Table 1-1
Estimated Annual Benefits and Costs of Air-Pollution Control in the United States

	Stationary Sources	Motor Vehicles	Total
Benefits			
Human health	$16.8	$0.2	$17.0
Vegetation	0.0	0.7	0.7
Materials	0.7	0.2	0.9
Soiling and aesthetics[a]	n.a.[b]	n.a.	2.8
Total benefits			$21.4
Costs			
Capital expenditures	4.2	3.5	7.7
Operating and maintenance[c]	3.4	4.6	8.0
Regulation and monitoring	n.a.	n.a.	0.2
Research and development	n.a.	n.a.	0.8
Total costs			$16.7

Source: Benefits: Adapted from A. Myrick Freeman III, *The Benefits of Air and Water Pollution Control: A Review and Synthesis of Recent Estimates* (U.S. Council on Environmental Quality, 1979). Costs: Adapted from Gary L. Rutledge, Pollution abatement and control expenditures in constant and current dollars, 1972-77, *Survey of Current Business* 59, no. 2 (1979):13-20; and Gary L. Rutledge and Betsy O'Connor, Capital expenditures by business for pollution abatement, 1977, 1978, and planned 1979, *Survey of Current Business* 59, no. 6 (1979):20-22.

Note: Benefits are for 1978 and costs are for 1977. Both are expressed in billions of 1978 dollars.

[a]Does not include visibility.

[b]n.a. indicates data not available.

[c]Includes depreciation. Does not include credit for recovery of product or byproducts.

Optimal Degree of Air-Pollution Control

The benefits of an air-pollution-control policy are the resulting reductions in air-pollution damages, and the costs are the resources necessary to implement the policy. Figure 1-1(a) illustrates typical relationships between the degree of control of an air pollutant and total benefits and costs. Both total benefits and total costs can be expected to increase as the percentage of the pollutant controlled increases, with total costs increasing at an increasing rate and total benefits increasing at a decreasing rate. The net benefits of control are equal to the difference between total benefits and total costs. In figure 1-1(a) the maximum net benefit is distance AB, and L^* is the optimal degree of control.

The optimal degree of control also can be determined from the increments to total benefits and total costs as the degree of control increases. The increment to total benefits when control increases by one unit is

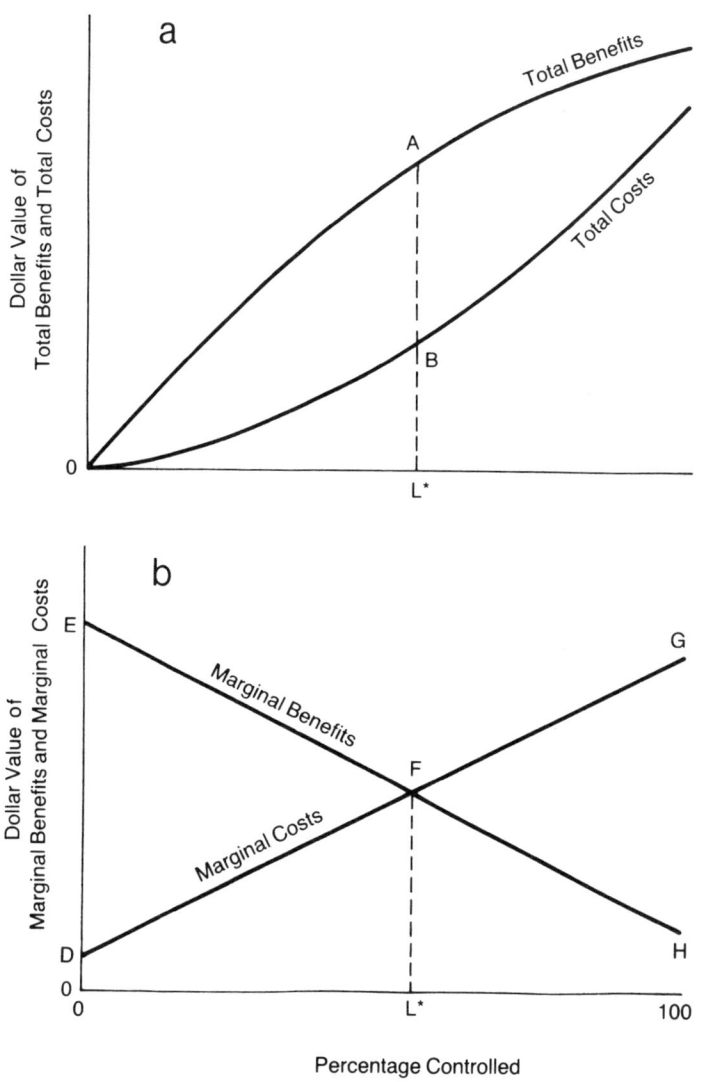

Figure 1-1. Optimal Degree of Air-Pollution Control

referred to as the *marginal benefit,* and the increment to total costs is referred to as the *marginal cost*. As shown in figure 1-1(b), marginal benefits are greater than marginal costs to the left of L* and are greater than marginal costs to the right of L*. Therefore, increases in control up to L* increase total benefits more than total costs, whereas increases in control beyond L* increase total costs more than total benefits. Thus the optimal degree of control is that for which marginal benefits are equal to marginal costs.[2]

For the case illustrated in figure 1-1, and in most actual pollution-control situations, the optimal degree of control is less than 100 percent because marginal costs increase, and marginal benefits decrease, as the degree of control increases. Thus attainment of the optimal degree of control does not imply the total elimination of air-pollution damages; and, conversely, the existence of air-pollution damages does not imply that the degree of control is less than optimal. Therefore, the observation that private markets result in substantial air-pollution damages does not in itself imply that government policies to increase the degree of air-pollution control are desirable. However, consideration of the way private markets operate does indicate that government action generally is required to attain the optimal degree of pollution control.

In considering the operation of private markets, it is convenient to think of pollution control as a productive activity the output of which is environmental quality. Like other productive activities, pollution control will be undertaken up to the point where the marginal benefit to the individual performing it is equal to his marginal cost. Therefore, if the benefits and costs accrued to the same individual, the optimal degree of pollution control would result.

The marginal benefits of a productive activity will accrue to the individual performing it if he uses the output himself, or if he sells the output for a price equal to its marginal benefit to the person using it. The latter condition will be satisfied for most types of output, because each consumer will be willing to pay a price equal to his marginal benefit. However, the output of pollution control generally can not be sold for a price equal to its marginal benefit because environmental quality is a *public good*. That is, each unit of environmental quality is available equally to all individuals in the relevant area whether they pay for it or not.[3]

Coase (1960) argued that private negotiations nevertheless would result in the optimal level of pollution control if contracts could be negotiated and enforced without cost (that is, if there were no transactions costs). Coase's argument can be illustrated using figure 1-1(b). Assume for simplicity that all damages of air pollution are incurred by individuals other than the polluter. Then, in the absence of negotiations, the percentage of control would be 0 because control would result in costs but no benefits for the polluter. However, the potential would exist for negotiations that made both the polluter and those damaged by the pollution better off, because marginal benefits from control (OE in figure 1-1(b)) would be greater than the marginal costs of control (OD). If those damaged by pollution paid the polluters an amount less than OE but greater than OD to undertake one unit of control, both would be better off. The potential for such mutually beneficial increases in pollution control exists up to the optimal degree of control, L*, at which point marginal benefits are equal to marginal costs.

Introduction

The total potential gain from increasing control from 0 percent to L* is the area EDF.[4]

The conclusion that the optimal degree of pollution control can be obtained through private negotiations has come to be known as the *Coase theorem*. Despite its importance in the development of the economic theory of pollution control, the Coase theorem provides little basis for complacency concerning most actual air-pollution-control situations. Because transactions costs are not in fact equal to zero, the type of negotiations he envisaged are likely to occur only when small numbers of individuals are involved.[5] Since most important pollution problems involve large numbers of individuals, the optimal degree of pollution control is unlikely to be attained by private actions.

It should also be emphasized that the term *optimal* in this context implies only that no potential mutually beneficial exchanges remain unexploited. Issues such as the equity or fairness of pollution-control decisions are not considered, although they may be very important in society's overall evaluation of the desirability of alternative outcomes. For example, a situation in which a poor person had to make substantial payments to a rich polluter in order to reduce the damage imposed on him by the polluter might be considered socially undesirable, even if it resulted in the "optimal" level of pollution.

The conclusion that private markets generally will not result in optimal air-pollution control implies that government intervention in this area has the potential to increase social well-being. However, for this potential to be realized it is necessary both that the appropriate degree of pollution control be determined and that appropriate policy instruments be chosen for achieving this degree of control. The choice of policy instruments has received considerable attention in the economics literature and will not be discussed here.[6] Instead, this book concentrates on the use of benefit-cost analysis to determine the optimal degree of control.

Benefit-Cost Analysis

Benefit-cost analysis of an air-pollution-control policy requires a prediction of the direct and indirect effects of the policy, expression of these effects in terms of common units, and a determination of the net impact on social well-being. The benefit-cost-analysis process is illustrated schematically in figure 1-2.

The first step in the analysis (labeled 1 in figure 1-2) is an evaluation of the effects of the policy on the decisions of the pollutant sources. For example, if the policy imposes a specific limitation on the emission rate of pollutants for a factory, it must be determined what control equipment or

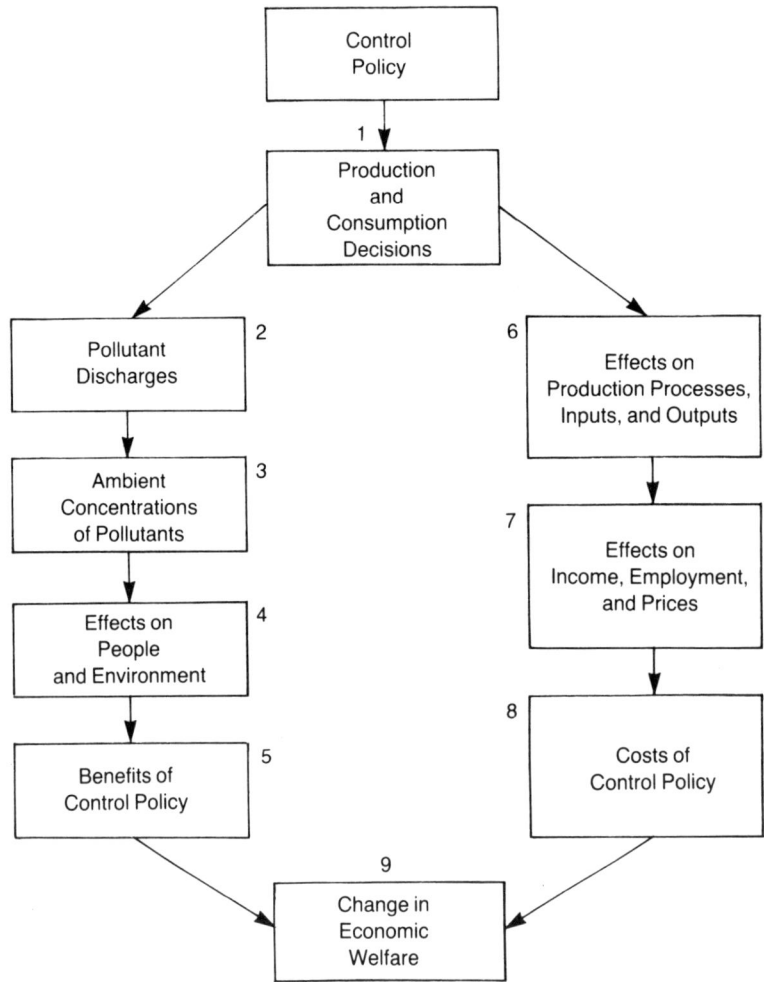

Figure 1-2. Steps in Benefit-Cost Analysis

changes in the production process will be chosen to satisfy the new requirement. If the policy affects the characteristics of a final product, such as automobiles, the effect on consumption decisions has to be determined.

The responses of pollutant sources to the control policy will affect both pollutant discharges and the use of resources. The effects on pollutant discharges give rise to the policy's benefits and are evaluated in steps 2 through 5. The effects on the use of resources give rise to the costs of the policy and are evaluated in steps 6 through 8. It should be noted that these

Introduction

definitions of benefits and costs are not synonymous with, respectively, positive and negative effects on social well-being. Thus some effects on pollutant discharges may be adverse (for example, if land or water pollution is increased as a result of an air-pollution policy) and some effects on resource use may increase well-being (for example, if the policy offsets a preexisting market distortion such as monopoly).

Analysis of the effects of production and consumption decisions on pollutant discharges (step 2) and of pollutant discharges on ambient concentrations (step 3) requires knowledge of the physical processes involved in the generation, transportation, dispersion, and transformation of the pollutants.

Analysis of the effects of the ambient concentrations on the exposed individuals and portions of the environment (step 4) requires knowledge both of physical and biological processes and of the economic responses to these effects. For example, determination of the effects of changes in pollutant concentrations on agricultural productivity requires knowledge of the effects of pollutants on the yields of various crops and of the responses of farmers in choosing which crops to plant. The failure adequately to account for economic responses has been a serious flaw in past studies of the effects of pollution.

Step 6, analysis of the effects of production and consumption decisions on production processes and the choices of inputs and outputs, requires procedures for estimating the resources required to implement these decisions. The effects of these changes on income, employment, and prices, evaluated in step 7, requires knowledge of the economic responses of the affected individuals and firms, including the possibility that a firm would prefer to shut down rather than modify its operations to satisfy the requirements of a policy.

The fifth and eight steps of a benefit-cost analysis involve the expression of the effects of a policy in terms of common units that reflect the value of these effects to the affected individuals. Although these values generally are expressed in dollar terms, this does not imply that a benefit-cost analysis is concerned only with the financial effects of a policy.

The final step in a benefit-cost analysis is the determination of the net impact of the benefits and costs on social well-being. This requires both a criterion for determining what constitutes an increase in social well-being, and procedures for aggregating effects that occur at different times, involve different degrees of uncertainty, and affect different individuals.

Several published benefit-cost analyses illustrate the application of the techniques discussed in this book to actual air-pollution-control problems. Finklea et al. (1975), Harrison (1975), and Schwing et al. (1980) have estimated the benefits and costs of motor-vehicle-emissions controls. Cohen, Fishelson, and Gardner (1974) examined the benefits and costs of regulating residential-heating fuels. North and Merkhofer (1975) and Cohen (1977) analyzed emissions regulations for electric-power plants. Loehman et al. (1979) considered the geographic and socioeconomic distribution of the receiving population. Mendelsohn (1980) introduced the location of the emission source as a control variable.

Criticisms of Benefit-Cost Analysis

The recent dramatic growth in interest in the use of benefit-cost analysis to evaluate public policies has generated a number of critical reviews of the theoretical basis and techniques for application of benefit-cost analysis.

The principal focus of a critique of benefit-cost analysis presented by a committee of the U.S. House of Representatives (1976a) is the difficulty inherent in estimating the direct effects of a policy and translating these into economic terms. The committee argued that such estimates seldom can be made with any accuracy because of the poor quality or nonexistence of objective measurements. This can provide substantial opportunities for the agency or analyst to bias the result by a subjective selection of critical values, which may not be apparent to a reviewer. The committee cited an example of a U.S. Department of Agriculture study that predicted that a U.S. Environmental Protection Agency (EPA) rule would result in 5,200 head of cattle being destroyed at a cost of $2.6 million. The rule was adopted, with 3 head eventually destroyed at a cost of $1,500. The committee further argued that analyses tend to emphasize the immediate, intended effects of the policy and to neglect longer-term undesirable or unintended effects that may be more difficult to quantify.[7]

Baram (1980) argued that "monetization of environmental and health amenities constitutes an inappropriate treatment of factors that transcend economics." He expressed concern for the evaluation of the distribution of the costs and benefits among social groups and concluded that distributional decisions cannot be made by "unaccountable analysts" who are outside the proper and traditional arenas for adjudicating the constitutional guarantees of due process, equal protection, and property rights.

The primary concerns of these and other critics may be summarized as follows:
1. Sufficiently great uncertainty exists in the estimates of the different components of the analysis that the results are frequently misleading and provide opportunities for mischief by less-than-objective analysts.
2. The selection of values for some variables and the analysis of distributional implications is best left to elected representatives and the courts rather than being buried within the results presented by the benefit-cost analyst.

Introduction

The first point clearly represents a real problem for benefit-cost analysis, but it is at least equally relevant for other possible ways of evaluating public policies. As Williams (1972) pointed out, a major advantage of benefit-cost analysis is that it provides a formal and explicit framework within which the effects of a policy can be compared, thereby making it possible to evaluate the effects of possible errors on the final decision. This virtue is not shared by such alternatives to benefit-cost analysis as simply leaving the decision to the political process. Similarly, it is true that value judgments on such issues as the desirability of the distributional consequences of a policy should not be implicitly imposed by the analyst; however, this is less of a problem for benefit-cost analyses than for less-explicit decision-making processes.

Plan of the Book

The economic theory underlying benefit-cost analysis is discussed in part I. Chapter 2 discusses the derivation of criteria for evaluating the net impact of a policy on social well-being. Chapters 3 and 4 discuss appropriate rules to use in aggregating benefits and costs over individuals and over time, respectively. Chapter 5 considers the evaluation of uncertain benefits and costs, and chapter 6 discusses the use of market data to assign dollar values to the effects of a policy.

Part II discusses the application of benefit-cost analysis to air-pollution-control problems. Chapter 7 discusses procedures for quantifying air-pollution effects. Chapters 8 through 11 discuss specific estimation procedures and existing empirical results for health, vegetation, materials, and aesthetics benefits, respectively. Chapters 12 and 13 discuss estimation procedures and existing empirical results for determination of control costs.

Appendix A discusses briefly the sources, measurement, and effects of the major air pollutants. Appendix B discusses the principal types of control equipment. Estimation results for order-of-magnitude cost functions are presented in appendix C.

Notes

1. However, the estimates do suggest that the total costs of existing control policies for motor vehicles are greater than the total benefits on a national scale.

2. In mathematical terms the marginal benefit is the derivative of total benefits with respect to the degree of control, and the marginal cost is the derivative of total cost. The condition that marginal benefit equal marginal

cost is thus the first-order condition for maximization of the difference between total benefits and total costs.

3. Samuelson (1955) discusses the distinguishing characteristics of public goods and the reasons that private markets generally will not result in the production of the optimal quantity of a public good.

4. Similarly, if no pollution could occur without the permission of those damaged by it, the degree of control in the absence of negotiations would be 100 percent. However, since the marginal cost of control would be greater than the marginal benefit, the potential for mutually beneficial reductions in control would exist until point L^* was reached.

5. An interesting example of negotiations occurring when a small number of individuals was affected, but not when many were, is cited by Baumol and Oates (1975). A Swedish refinery discharged corrosive materials into the air when lower-quality petroleum was refined. A nearby automobile plant negotiated with the refinery, which agreed to undertake the corrosive activities only when the wind was blowing away from the automobile plant and toward a large number of households, who did not participate in the negotiations.

6. Economists have tended to favor the use of incentives, such as pollution taxes, rather than direct regulations. See, for example, Baumol and Oates (1975) and Kneese and Schultze (1975). Some of the complications involved in the use of such incentives are discussed by Rose-Ackerman (1973) and Tietenberg (1978). Weitzman (1974) discusses the possible superiority of direct regulation under conditions of uncertainty.

7. When very long-term effects are involved, intergenerational equity becomes an important consideration. This issue is discussed by Mishan (1977) and Page (1977).

Part I
The Theory of Benefit-Cost Analysis

2 Theoretical Foundations of Benefit-Cost Analysis

The objective of a benefit-cost analysis of a public policy, such as an air-pollution-control policy, is to determine whether the policy will contribute to the well-being of society. This requires, first, a definition of social well-being and, second, a practical criterion for determining whether the effects of a policy result in an increase in the well-being of society.

This chapter discusses the derivation of criteria for evaluating policies using the economic theory of social-welfare maximization.[1] The purpose is to make explicit the value judgments and informational requirements underlying each of the major criteria that have been proposed for use in benefit-cost analysis.

Social Welfare

The basic goal of society is assumed to be the maximization of social welfare, which is specified to be a function of the well-being of each of the individuals in society. This can be expressed mathematically as

$$W = W(U_1, U_2, \ldots, U_n) \qquad (2.1)$$

where W is social welfare and U_j is the well-being of individual j. Economists have traditionally used the word *utility* to describe the well-being of an individual, and this terminology will be used here.

The individualistic definition of social welfare explicit in equation 2.1 presumably would receive general support in democratic societies, but it is not the only possible definition. For example, the organic theory of the state holds that what is important is the welfare of the state per se, rather than the well-being of individuals. Thus, even at this level of abstraction, the economic analysis of social welfare cannot claim to be entirely objective.

The relationship between changes in individual utility and social welfare can be examined using the total differential of equation 2.1

$$dW = \frac{\partial W}{\partial U_1} dU_1 + \frac{\partial W}{\partial U_2} dU_2 + \ldots + \frac{\partial W}{\partial U_1} \frac{\partial U_1}{\partial U_2} dU_2 + \ldots \qquad (2.2)$$

The calculus notation used in equation 2.2 and subsequent equations can be interpreted as follows. The letter d indicates a small change in a variable; for example, dU_1 indicates a small change in the utility of individual 1. The term $\partial W/\partial U_1$ indicates the effect on W of a small change in U_1 when all other things that affect W are held constant. Multiplying $\partial W/\partial U_1$ by dU_1 thus tells us the change in welfare that would result from a given (small) change in U_1. Adding together the effects of the changes in utility for all individuals yields the total change in social welfare, dW.

Second-order terms such as $(\partial W/\partial U_1)(\partial U_1/\partial U_2)dU_2$ indicate the effect on social welfare of a change in an individual's utility that results from a change in some other individual's utility (for example, as a result of feelings of envy or empathy). Following the usual practice in economics, we will omit second-order (and higher) terms in considering the effects of policies on social welfare. This can be viewed either as an empirical judgment that the effects on each individual's utility of changes in the well-being of others are sufficiently small that they can be ignored, or as a value judgment that such effects, if they exist, should not affect the evaluation of public policies. With this modification, equation 2.2 becomes

$$dW = \frac{\partial W}{\partial U_1} dU_1 + \frac{\partial W}{\partial U_2} dU_2 + \ldots + \frac{\partial W}{\partial U_n} dU_n \qquad (2.3)$$

The economic theory of social welfare also assumes that an increase in the utility of any individual, other things constant, will result in an increase in social welfare [that is, $(\partial W/\partial U_j) > 0$ for all j]. Again, this can be viewed either as an empirical judgment or as a value judgment. Given this assumption, it can be concluded from equation 2.3 that any policy resulting in an increase in utility for one or more individuals, without decreasing the utility of anyone else, would increase social welfare. This is the basis for the definition of *economic efficiency* (also referred to as Pareto optimality):

Economic efficiency is attained only if it is not possible to make anyone better off without making someone else worse off.

A situation is defined as *inefficient* if this condition is not satisfied, that is, if it is possible to make someone better off without making anyone else worse off.

Economic-Efficiency Criterion

Social welfare is not maximized unless all changes that increase it are undertaken. Therefore, the attainment of economic efficiency is a necessary con-

Theoretical Foundations

dition for social-welfare maximization. Since, additionally, the concept of economic efficiency is based on widely accepted value judgments and requires information only on the direction of change in individual utilities, it might appear to provide an attractive theoretical foundation for benefit-cost analysis.

However, economic efficiency has two basic shortcomings as a criterion for evaluating public policies. First, although economic efficiency is necessary for social-welfare maximization, it is not sufficient. There are an infinite number of efficient situations, and no basis is provided for choosing among them. Second, an economic-efficiency criterion cannot be used to evaluate a policy that results in losses to anyone, even if the policy moves the economy from an inefficient situation to an efficient one.

The concept of economic efficiency, and its shortcomings as a theoretical foundation for benefit-cost analysis, are illustrated in figure 2-1 for the case of a society consisting of two individuals, Smith and Jones. Changes in the utilities of Smith and Jones are indicated by movements along the vertical and horizontal axes, respectively. Different points in the diagram reflect the utility received by each as a result of alternative decisions by society with respect to the commodities produced, their distribution between the two individuals, and any other factors that affect the levels of well-being attained by the individuals.

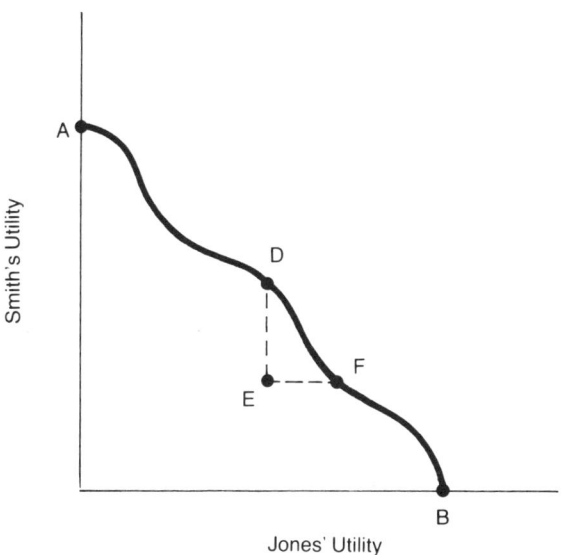

Figure 2-1. Utility-Possibility Frontier

The points on the wavy line between A and B show the maximum attainable utility for each individual given the utility of the other. Since the line shows the maximum possible combinations of utility, it is referred to as the *utility-possibility frontier*. If the utilities of the individuals corresponded to a point on the frontier, it would not be possible to make one better off without making the other worse off. For example, a move from point F to point D would make Smith better off but Jones worse off. Therefore, points on the utility-possibility frontier correspond to situations in which economic efficiency is attained. There are an infinite number of efficient situations, and the efficiency criterion provides no basis for choosing between them.

If the utilities of the individuals corresponded to a point within the utility-possibility frontier, then gains by one individual could occur without losses to the other. For example, if the utilities of the individuals corresponded to point E, a policy that resulted in a move to point F would result in an increase in Jones's utility with no decrease in Smith's, whereas a move to D would increase Smith's utility with no decrease in Jones's. A move anywhere in the area DEF would result in increases in the utility of both Smith and Jones.

Although the potential exists for making some individuals better off without making anyone else worse off whenever the initial situation is not efficient, this does not imply that a policy that moves the economy from an inefficient situation to an efficient one would itself necessarily satisfy an economic-efficiency criterion. For example, a policy that moved the economy from point E to (efficient) point B would make Jones better off but would make Smith worse off.

The efficiency criterion's inability to indicate which of the infinite number of efficient outcomes maximizes social welfare is important if society cares about the final levels of utility attained by its members. Since efficient situations include cases in which almost all well-being is experienced by only one individual (for example, points A and B in figure 2-1), as well as cases in which well-being is more evenly distributed (for example, points D and F), society might well prefer one efficient solution to another.

If society does care about the distribution of well-being, benefit-cost analyses based on the assumption that it does not may be of no practical use in making decisions reflecting society's preferences. On the other hand, as discussed later in this chapter, any criterion that attempts to incorporate society's preferences concerning the distribution of well-being must face the difficult task of determining just what those preferences are.

The efficiency criterion's inability to evaluate policies that result in losses to anyone is its most serious shortcoming. Virtually all policies actually considered by government result in losses to some individuals. For example, an air-pollution-control policy that required the installation of pollution-control equipment by a factory would result in a loss of profits to the owners of the factory, and hence could not be evaluated by the efficiency criterion. This problem could be avoided if the owners were compensated for their losses, with the funds for the compensation obtained by taxing away part of the benefits to the individuals who benefited from the reduction in pollution. However, unless the benefits for each individual could be measured quite precisely, the tax payment for some individual might exceed his benefits, so that the policy would result in a net loss to that individual. Even if this problem was avoided, the changed production decisions of the firm would be likely to result in losses to one or more of its suppliers.

Social-Welfare Criterion

Because the efficiency criterion is too restrictive to be of much use in evaluating public policies, welfare economists have sought to develop criteria that allow for the possibility of policies that result in losses to some individuals nevertheless resulting in net gains to society as a whole. One possibility would be to use equation 2.3 directly. This will be referred to as the *social-welfare criterion*.

Equation 2.3 says that a policy results in an increase in social welfare when the weighted sum of the changes in individual utilities is positive. The weights, $\partial W/\partial U_j$, can be interpreted as the *marginal social significance* of each individual because they indicate the effect on social welfare of a small change in that individual's well-being.

Although equation 2.3 is strictly applicable only for policies resulting in small changes in utilities, the social-welfare criterion is readily generalized to allow for large changes in utility, as illustrated in figure 2-2. Efficient situations are again represented by points on the utility-possibility frontier. Society's preferences concerning the distribution of well-being are represented by the family of curves, U_0U_0, U_1U_1, U_2U_2, and so on. Each curve shows all combinations of individual utilities that yield a given level of social welfare. Since movements along a curve leave the level of social welfare unchanged, the slope of a curve shows the amount by which one individual's utility must increase to leave social welfare unchanged when the other individual's utility decreases. Therefore, the slope is equal to (the negative of) the ratio of the marginal social significances of the individuals.

The social-welfare criterion offers several potential advantages over the efficiency criterion. First, it provides a basis for choosing between different efficient situations. For example, point F is preferred to point G because it

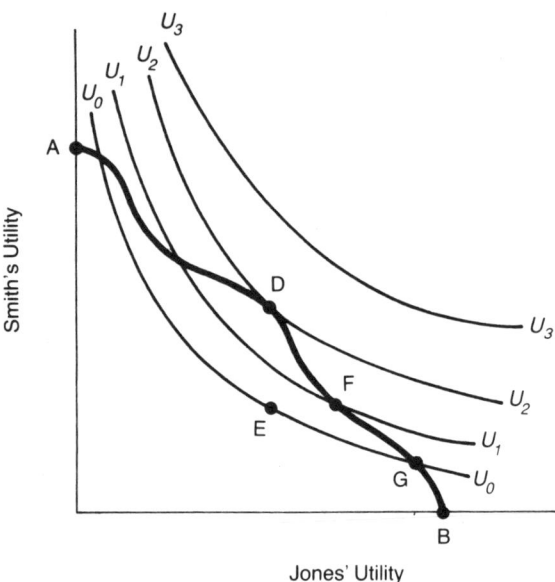

Figure 2-2. Maximization of Social Welfare

results in greater social welfare. Point D in turn is preferred to point F and in fact is the most-preferred point since it results in the greatest attainable social welfare.

The social-welfare criterion also permits the comparison of moves that result in an increase in one individual's utility and a decrease in the other's, whether the move is from an inefficient point such as E to an efficient one such as G, or is a move between two efficient points, as is the move from point D to point F. The diagram also illustrates that an inefficient point may be as desirable socially as an efficient one, as in the case of points E and G, and may be preferred to an efficient point, as, for example, E is preferred to B.

Therefore, the social-welfare criterion overcomes the major shortcomings of the efficiency criterion. However, implementation of the social-welfare criterion requires far more information. Unlike the efficiency criterion, which requires only the assumption that each individual counts positively, the social-welfare criterion requires knowledge of the actual value of the marginal social significance of each individual. The relevant values are those of society as a whole. Whether or not a democratic society is capable of specifying these values remains an open question in economic theory.[2]

Theoretical Foundations

Application of the social-welfare criterion also requires more information concerning individual utilities. The efficiency criterion requires knowledge only of the direction of change in individual utilities, whereas the social-welfare criterion requires interpersonally comparable measures of the magnitude of these changes. Unfortunately, there is no scientific way of deriving such measures.[3]

Economic-Welfare Criterion

The efficiency criterion and the social-welfare criterion represent two extremes with respect to required information and generality of application. The efficiency criterion requires minimal assumptions with respect to the marginal social significance and changes in utility for each individual, but it is too restrictive to be of practical use in evaluating policies; the social-welfare criterion, on the other hand, is minimally restrictive but requires too much information to be implemented. To be useful, a theoretical foundation for benefit-cost analysis must be less restrictive than the efficiency criterion but must also require less information than the social-welfare criterion.

One way to reduce the amount of information required is to narrow the scope of the analysis to include only the effects of alternative policies on the production and distribution of goods and services. Other possible effects on well-being, such as impacts on personal freedom or due process, could be considered to be outside the scope of analysis. In other words, the evaluation of policies could be restricted to their effect on economic welfare, rather than on social welfare. At the current level of abstraction this distinction may be more apparent than real, because goods and services can be defined as broadly as desired. However, to the extent that possible influences on welfare are omitted, informational requirements will be reduced at the expense of a less-general analysis.[4]

Informational requirements can be reduced further by assuming that each individual's utility depends only on his own consumption of goods and services. Because goods and services can be defined to include common-property resources such as clean air, this assumption does not rule out externalities in consumption of the type that underlie pollution problems. However, it does rule out factors such as paternalism, moral judgments, and so forth. This restriction can be viewed either as an empirical judgment that the effects of one individual's consumption on others' utilities are sufficiently small that they can be ignored, or as a value judgment that such effects, if they exist, should not affect the evaluation of public policies.

Given these restrictions, equation 2.1 can be written

$$W = W[U_1(X_{11}, \ldots, X_{1m}), \ldots U_n(X_{n1}, \ldots, X_{nm})] \qquad (2.4)$$

where W now represents economic welfare and X_{jk} is individual j's consumption of good k. Equation 2.3 then becomes

$$dW = \frac{\partial W}{\partial U_1} \frac{\partial U_1}{\partial X_{11}} dX_{11} + \ldots + \frac{\partial W}{\partial U_n} \frac{\partial U_n}{\partial X_{nm}} dX_{nm} \qquad (2.5)$$

or

$$dW = \sum_{j=1}^{n} \sum_{k=1}^{m} \frac{\partial W}{\partial U_j} \frac{\partial U_j}{\partial X_{jk}} dX_{jk} \qquad (2.6)$$

where dX_{jk} is the change in individual j's consumption of good k and $\partial U_j/\partial X_{jk}$ is the marginal utility of good k to individual j.

A policy will be said to satisfy the economic-welfare criterion if the value of dW calculated using equation 2.6 (or its generalization for the case of nonmarginal changes) is positive.

Potential-Compensation Criteria

The information required to evaluate policies using the economic-welfare criterion is less extensive than that required in using the social-welfare criterion, but information on the marginal social significance of each individual is still needed, as are interpersonally comparable measures of marginal utilities. The attempt to find a criterion that avoids the need for this information in evaluating public policies has been characterized as the "new welfare economics."[5]

Kaldor (1939) and Hicks (1939) independently suggested similar criteria that require substantially less information than the economic-welfare criterion. Both criteria are based on the concept of potential compensation. The criterion suggested by Kaldor is:

A policy is desirable if the gainers from the policy could more than compensate the losers if the policy were adopted,

whereas the criterion suggested by Hicks is:

A policy is desirable if the losers could not compensate the gainers for foregoing the policy.

Theoretical Foundations

Both criteria are based on dollar measures of gains and losses, and neither requires that the compensation actually occur. They differ in that Kaldor's criterion is based on calculations of potential compensation using the prices that exist before the policy is enacted, whereas Hicks's criterion is calculated using the prices that would exist if the policy were carried out. Scitovsky (1941) pointed out that, if the policy in question resulted in large changes in prices, it was possible that either criterion could indicate first that a policy was desirable and, after the policy was enacted, that a reversal to the original situation was desirable. Therefore, he proposed a double criterion incorporating both Kaldor's criterion and Hicks's criterion.

The criteria proposed by Kaldor, Hicks, and Scitovsky are all based on the concept of potential compensation and will be referred to as *potential-compensation criteria*. If no changes in prices occur as a result of the policy, as will be assumed for the remainder of this chapter,[6] then the three potential-compensation criteria would reach identical conclusions with respect to the desirability of a given policy.

The potential-compensation criteria say that a policy is desirable if the sum of the dollar values of the effects of the policy to gainers and to losers is positive. When prices are constant, the dollar value to an individual of a unit change in his consumption of a good (or service) will be equal to its price. Therefore, the total dollar value of the gain or loss that individual j incurs as a result of the policy (that is, his net benefit), is

$$\sum_k P_k \, dX_{jk}$$

where P_k is the price of good k and dX_{jk} is the change in individual j's consumption of good k resulting from the policy. The potential-compensation criteria say that the desirability of a policy can be evaluated by summing these dollar values across all individuals,[7] that is, by calculating

$$\sum_j \sum_k P_k dX_{jk} \qquad (2.7)$$

Comparison of Potential-Compensation and Economic-Welfare Criteria

We can investigate the conditions under which the potential-compensation criteria and the economic-welfare criterion would result in the same conclusions about the desirability of a proposed policy by expressing the economic-welfare criterion in terms of the prices and quantities of goods and services. The economic-welfare criterion, equation 2.6, says that the desirability of a policy should be evaluated using the formula

$$dW = \sum_j \sum_k \frac{\partial W}{\partial U_j} \frac{\partial U_j}{\partial X_{jk}} dX_{jk} \qquad (2.8)$$

If all individuals face the same prices for goods, and arrange their purchases so as to maximize their utility, then

$$\frac{\partial U_j}{\partial X_{jk}} = \frac{\partial U_j}{\partial Y_j} P_k \qquad (2.9)$$

where $\partial U_j/\partial Y_j$ is the marginal utility of income to individual j.[8] Substituting 2.9 into 2.8, the economic-welfare formula can be written

$$dW = \sum_j \sum_k \frac{\partial W}{\partial U_j} \frac{\partial U_j}{\partial Y_j} P_k dX_{jk} \qquad (2.10)$$

Formulas 2.7 and 2.10 differ in the term

$$\frac{\partial W}{\partial U_j} \frac{\partial U_j}{\partial Y_j} \qquad (2.11)$$

which appears in 2.10 but not in 2.7. This term, which can be written as $\partial W/\partial Y_j$, indicates the effect on society's economic welfare of a small change in individual j's income and will be referred to as the *marginal social utility* of individual j's income.

The marginal social utilities of income can be normalized with no loss of generality be defining

$$\alpha_j = \frac{\partial W/\partial Y_j}{\partial W/\partial Y_{50}} \qquad j = 1, \ldots, n \qquad (2.12)$$

where $\partial W/\partial Y_{50}$ is the marginal social utility of income for the median individual.[9] Substituting 2.12 into 2.10 allows us to determine the (normalized) change in economic welfare resulting from a given policy as

$$\frac{dW}{\partial W/\partial Y_{50}} = \sum_j \sum_k \alpha_j P_k dX_{jk} \qquad (2.13)$$

The right-hand side of equation 2.13 can be rewritten as

$$\sum_j \sum_k P_k dX_{jk} + \sum_j \sum_k (\alpha_j - 1) P_k dX_{jk} \qquad (2.14)$$

Thus the effect of a policy on economic welfare can be expressed as the sum of its effect measured by a potential-compensation criterion:

Theoretical Foundations

$$\sum_j \sum_k P_k dX_{jk} \qquad (2.15)$$

and a term representing purely distributional effects:

$$\sum_j \sum_k (\alpha_j - 1) P_k dX_{jk} \qquad (2.16)$$

If the marginal social utility of income were equal for all individuals, α_j would be equal to 1 for all j, and the normalized effect of a policy on economic welfare, formula 2.14, would be equal to its effect as measured by a potential compensation criterion, formula 2.15. Therefore, the ranking of policies obtained using the economic welfare and potential compensation criteria would be identical. However, if marginal social utitilies of income are not equal, the ranking of policies obtained using formula 2.15 will be the same as that obtained using formula 2.14 only if the relative distribution across individuals of the effects of the policies on the consumption of goods are identical for all the policies in question *and* either the signs of 2.15 and 2.16 are the same, or the absolute value of 2.15 is greater than that of 2.16. Since these highly restrictive conditions will not be satisfied for most policies, the ranking of alternative policies obtained using a potential-compensation criterion cannot be expected to reflect their relative contribution to economic welfare unless the marginal social utilities of income are equal.

Inspection of expression 2.11 shows that the marginal social utility of income will not be equal for all individuals if either the marginal social significances, $\partial W / \partial U_j$, or the individual marginal utilities of income, $\partial U_j / \partial Y_j$, vary across individuals (unless any variations in these two terms happen to be exactly offsetting for all individuals). Marginal social significances represent society's judgments on the importance of the wellbeing of each individual. Whether or not they are equal for all individuals is an ethical question that is not in the province of economics. However, individual marginal utilities of income are reflected in income and consumption data and can be investigated empirically. The results of econometric studies of this issue indicate that individual marginal utilities of income are not equal. For example, Phlips (1974) and Theil (1975) found that the marginal utility of income decreases as an individual's income increases. Therefore, the marginal social utility of income cannot be assumed equal across individuals; and a potential-compensation criterion will not, in general, be equivalent to the economic-welfare criterion.

Potential Economic Welfare

Kaldor (1939) recognized that satisfaction of a potential-compensation criterion did not in general imply that a policy increased economic welfare.

However, he argued that a potential-compensation criterion provided a means of determining whether or not a policy increased *potential* economic welfare. The question of whether or not it increased the *actual* level of welfare, he felt, lay outside the appropriate domain of economic analysis. Thus, in advocating the use of a potential-compensation criterion, Kaldor was in essence recommending a division of labor between economists, who were to measure the impact of policies on potential economic welfare, and some (unspecified) noneconomists, who were to decide which, if any, redistributions were required to convert increases in potential welfare into increases in actual welfare.

Kaldor's proposal can be illustrated by a numerical example. Suppose a proposed policy would result in a net benefit of $220 to individual 1, for whom α_j is equal to 0.9, and a net benefit of $-\$200$ to individual 2, for whom α_j is equal 1.1. As shown in part I of table 2-1, the policy would satisfy Kaldor's potential-compensation criterion but, if no redistribution occurred, would result in a decrease in economic welfare. As shown in part II of the table, if $110 of individual 1's net benefit were redistributed to individual 2, the policy, together with this redistribution, would leave economic welfare unchanged. Therefore, economic welfare would be increased if the policy were undertaken and if a redistribution of more than $110 from individual 1 to individual 2 occurred.

Translation of the increase in potential economic welfare into an

Table 2-1
Illustration of Potential-Compensation Criterion

Individual	Marginal Social Utility of Income α_j	Net Benefit $\Sigma_k P_k dX_{jk}$	Distributional Effect $\Sigma_k (\alpha_j - 1) P_k \times dX_{jk}$	Change in Economic Welfare
I. Policy With No Redistribution				
1	0.9	$220	$-22	
2	1.1	-200	-20	
Σ_j		$ 20	$-42	$-22
II. Policy With Redistribution of $110				
Individual	Marginal Social Utility of Income	Net Benefit $-\$110$	Distributional Effect	Change in Economic Welfare
1	0.9	$110	$-11	
2	1.1	- 90	$- 9	
Σ_j		$ 20	-20	0

increase in actual economic welfare requires that sufficient redistribution actually take place. A possible objection to Kaldor's proposal is that redistribution might not occur because of political constraints. However, this assumes that society is able to determine that redistribution of the net benefits of a policy is desirable but is unable to ensure that the political system will implement the redistribution. Since it is unclear how society can express distributional preferences except through the political system, the failure of a redistribution to attain political support can be considered prima facie evidence that society does not consider it desirable.

However, even if all desirable redistributions take place, Kaldor's proposed division of labor might not result in the maximization of economic welfare because it implicitly assumes that costless methods of redistribution are available. Unfortunately, all feasible methods of redistribution involve both administrative costs and adverse incentive effects. For example, income taxes distort individuals' decisions concerning their allocation of time between work and leisure because the rewards of work are taxed, whereas those of leisure are not. As a result of such distortions, the total cost to the taxpayer is greater than the amount of revenue raised. Browning (1976) has estimated that raising a marginal dollar of revenue through taxes on labor income in the United States costs from $1.09 to $1.16 and that the cost of raising revenue through excise taxes may be substantially higher.

Because there are no costless methods of redistribution, satisfaction of a potential-compensation criterion need not imply that a policy would make it possible to increase economic welfare. To illustrate, if the costs of redistribution were equal to 10 percent of the amount transferred, a policy with the net benefits shown in table 2-1 could not result in an increase in economic welfare. If total costs equal to his full net benefits of $220 were imposed on individual 1, the amount redistributed to individual 2, net of the costs of redistribution, would be $200. The net benefits after redistribution would be zero for both individuals, and economic welfare would be unchanged by the policy. If only part of individual 1's net gains from the policy were redistributed, economic welfare would be decreased by the policy.

Summary

Consideration of the theoretical foundations of benefit-cost analysis reveals the value judgments and informational requirements that underlie alternative criteria for evaluating the contribution of policies to society's goals. The sobering conclusion is that there is no fully adequate criterion for evaluating proposed policies. The efficiency criterion involves the fewest value judgments and requires information only on the direction of change in the well-being of individuals, but it is too restrictive for practical application in evaluating policies. The social-welfare criterion is conceptually complete and provides theoretically correct measures of the contribution of policies to social welfare, but its informational requirements prevent its application.

The economic-welfare criterion requires somewhat less information than the social-welfare criterion, at the expense of being less conceptually complete. Despite the reduction in information required, the economic-welfare criterion is also unlikely to be directly applicable in evaluating policies because it requires information on the marginal social significance of each individual as well as interpersonally comparable measures of utility.

Potential-compensation criteria are less restrictive than the efficiency criterion and much easier to implement than the economic-welfare criterion. However, unless the marginal social utility of income is equal for all individuals, satisfaction of a potential-compensation criterion does not imply that a policy increases *actual* economic welfare. Furthermore, in the absence of costless methods of income redistribution, satisfaction of a potential-compensation criterion need not imply that a policy increases even *potential* economic welfare.

Of the available alternatives, the potential-compensation criteria appear the most promising, provided that a satisfactory method can be found for dealing with distributional issues. The next chapter discusses the procedures that have been proposed for resolving the distributional issues involved in benefit-cost analysis.

Notes

1. The economic theory of social-welfare maximization was first stated in its modern form by Bergson (1938). Bator (1957) provides an excellent graphical exposition. Henderson and Quandt (1971, pp. 254-292) provide a general survey using calculus.

2. The seminal work indicating that a democratic society cannot in general specify a social-welfare function is Arrow (1950). For a recent counterargument see Bailey (1979).

3. In technical terms, the efficiency criterion requires only ordinal measures of utility, whereas the social-welfare criterion requires cardinal measures. For a discussion of these two concepts of utility measurement, see Baumol (1977, pp. 190-195, 421-424).

4. For an interesting discussion of the potential consequences of omitting some possible influences on social welfare from the analysis, see Culyer (1977).

Theoretical Foundations

5. For a comprehensive (and highly technical) survey of the new welfare economics see Chipman and Moore (1978).

6. The issues that arise when prices do not remain constant are discussed in chapter 6.

7. For the case considered, in which all individuals are assumed to face the same prices, the formula simplifies to $\Sigma_k P_k dX_k$. Thus information is required only on the changes in the total quantities of goods, not on the changes in each individual's consumption of each good.

8. Equation 2.9 is one of the first-order conditions for utility maximizaton. See, for example, Henderson and Quandt (1971, pp. 18-19).

9. The discussion in this paragraph and the next is based on Azzi and Cox (1973).

3
Aggregation of Benefits and Costs over Individuals

Generally, benefits and costs of air-pollution-control policies will not be distributed uniformly across individuals. For example, a policy requiring the installation of control equipment at an electric-power plant may result in positive net benefits for individuals living near the plant and in net losses for the owners of the plant. Therefore, procedures are required for aggregating benefits and costs over individuals to determine the net effect of a policy on social well-being.

As discussed in chapter 2, the net effect of a policy on economic welfare can be expressed as

$$\sum_j \sum_k \alpha_j P_k dX_{jk} \tag{3.1}$$

where α_j is the (normalized) marginal social utility of income for individual j, dX_{jk} is the change in individual j's consumption of commodity k, and P_k is the price of commodity k. Formula 3.1 can be rewritten as

$$\sum_j \alpha_j \sum_k P_k dX_{jk} \tag{3.2}$$

Since $\sum_k P_k dX_{jk}$ is the dollar value to individual j of his gain or loss, formula 3.2 shows that the net effect of a policy on economic welfare is equal to the weighted sum of individual gains and losses, using marginal social utilities of income as distributional weights.

If complete information were available on the net gain or loss for each individual, as well as on his marginal social utility of income, then formula 3.1 or 3.2 could be applied directly. However, this procedure is not feasible in practice because complete data on the distribution of gains and losses is prohibitively expensive to obtain, and the values of marginal social utilities of income cannot be directly observed.[1]

This chapter discusses several procedures that have been proposed for reducing the amount of information required in aggregating benefits and costs over individuals. Some of the proposed procedures reduce the amount of required information by avoiding the use of explicit distributional weights. These proposals include simply ignoring the distribution of benefits and costs, providing information on the distributional effects of a policy but not trying to aggregate them, and using aggregation rules that

29

require information only on the direction of change in the marginal social utility of income. The second approach is to design simplified techniques for using marginal social utilities of income as explicit distributional weights.

Aggregation Without Explicit Distributional Weights

The simplest, and most commonly used, procedure for aggregating over individuals is to calculate the total dollar value of net benefits without regard to whom they accrue. This procedure, which is equivalent to using a potential-compensation criterion as the sole basis for evaluating policies, has the advantage of not requiring information on the distribution of net benefits or on the marginal social utility of income. However, as discussed in chapter 2, the ranking of policies obtained using a potential-compensation criterion will not in general reflect their relative contribution to economic welfare unless the marginal social utility of income is equal for all individuals. Furthermore, in the absence of costless methods of income redistribution, satisfaction of a potential-compensation criterion need not even imply that a policy increases potential economic welfare.[2]

McKean (1958) recognized the undesirability of basing the choice of policies solely on the total dollar value of their net benefits. He recommended the addition of exhibits showing the distribution of net benefits so that the social decision maker could take distributional considerations into account. This is clearly a step in the right direction, but one that leaves unsolved the difficult problem of how the data on the distribution of net benefits should be evaluated. The amount of raw data supplied to the decision maker would be substantial, and without an explicit aggregation procedure it would be difficult to attain consistency from decision to decision. Also, if the aggregation procedure used is not made explicit, no basis is provided for the public to review the distributional weights underlying the final choice of policies.

Willig and Bailey (1981) have attempted to derive practical aggregation procedures based on what they consider to be widely accepted assumptions concerning the marginal social utility of income. They assumed that the marginal social utility of income for an individual depends only on his income, decreases as income increases, and is non-negative for all individuals. A policy is then said to *social-welfare dominate* another if it would be preferred by any social decision maker who accepted these assumptions about the marginal social utility of income.

Willig and Bailey showed that policy A social-welfare dominates policy B if and only if the total net benefits of the poorest h individuals under

Aggregation over Individuals

policy A are greater than under policy B, for all values of h. That is, labeling the individuals so that individual 1 is poorer than 2, who is poorer than 3, and so on, policy A social-welfare dominates policy B if and only if

$$\sum_{j=1}^{h} NB_j^A > \sum_{j=1}^{h} NB_j^B \qquad h = 1, 2, \ldots, n \qquad (3.3)$$

where NB_j^A is the dollar value of the net benefits of policy A to individual j, NB_j^B is the net benefits of policy B to individual j, and there are a total of n individuals.

When $h = n$, the condition represented in inequality 3.3 is identical to a potential-compensation criterion's requirement that the sum across all individuals of the dollar values of gains and losses under policy A be greater than under policy B. Thus a necessary condition for policy A to social-welfare dominate policy B is that A be preferred to B using a potential-compensation criterion. Added to this condition are the conditions that the sum of net benefits under A be greater than under B for all other groups of individuals, ranked by income.[3]

Willig and Bailey proposed that the conditions in 3.3 be used as the basis for evaluating public policies.[4] The principal advantage of this procedure is that the actual value of the marginal social utility of income for each individual does not have to be determined. The principal disadvantage is that only a partial ordering of policies is obtained. Any policy satisfying the conditions in 3.3 would also be chosen using the economic-welfare criterion, formula 3.1 or formula 3.2; but some policies that would be chosen using the economic-welfare criterion would not be chosen using the conditions in 3.3.

To illustrate this disadvantage of the Willig and Bailey procedure, let policy A be the policy whose effect on economic welfare is calculated in table 3-1 and policy B be the policy of maintaining the status quo. Policy A increases economic welfare and therefore would be chosen over policy B using the economic-welfare criterion. However, if individual 1 were poorer than individual 2 (consistent with the higher value of α for individual 1), the conditions in 3.3 would not be satisfied because the sum of net benefits for policy A when $h = 1$ (-200) is less than for policy B (0).

The Willig and Bailey procedure is also subject to criticism concerning their assumptions about the marginal social utility of income. In particular, the assumption that the marginal social utility of income for an individual depends only on his income omits from consideration factors such as age, race, medical needs, and so forth, which society may consider relevant. Also, the assumption that the marginal social utility of income decreases as income increases is widely, but not universally, accepted.

Table 3-1
Effect on Economic Welfare of Policy A

Individual	Marginal Social Utility of Income α_j	Net Benefit $\Sigma_k P_k dX_{jk}$	Distributional Effect $\Sigma_k (\alpha_j - 1) P_k \times dX_{jk}$	Change in Economic Welfare
1	1.2	−200	−40	
2	0.9	280	−28	
Total		80	−68	12

Aggregation with Explicit Distributional Weights

Feldstein (1974a) has proposed a procedure for explicitly incorporating distributional weights in benefit-cost analyses. He suggested that policies be evaluated in terms of the *uniformly distributed dollar* (UDD) equivalent of their net benefits, where a UDD is a single dollar divided equally among all individuals in the nation. The welfare value of a UDD is

$$\sum_Y \alpha(Y) f(Y), \qquad (3.4)$$

where $f(Y)$ is the fraction of individuals with income Y and $\alpha(Y)$ is the normalized marginal social utility of income for an individual with income Y.[5]

The average net benefit of a policy to an individual with income Y is represented by $w(Y)$. The value of a policy, measured in UDDs, is then

$$\frac{N \sum_Y w(Y) \alpha(Y) f(Y)}{\sum_Y \alpha(Y) f(Y)} \qquad (3.5)$$

where N is the total number of individuals.

To make this procedure useful in practice, Feldstein suggested that approximation formulas be used for the distribution of benefits and costs and for the distributional weights, α. The formula suggested for α is

$$\alpha(Y) = \left(\frac{Y}{Y_{50}}\right)^{-\delta} \qquad (3.6)$$

where Y_{50} is the median level of income. The parameter δ in equation 3.6 measures the responsiveness of the marginal social utility of income to

Aggregation over Individuals

income. If $\delta = 0$, $\alpha(Y) = 1.0$ for all Y. Substituting in 3.4, this implies the welfare value of a uniformly distributed dollar is 1.0; and, from 3.5, the value of a policy is then equal to the unweighted sum of its net benefits. For values of $\delta > 0$, the marginal social utility of income decreases as income increases. Therefore, when $\delta > 0$, the value of a policy will be greater than the unweighted sum of its net benefits if the distribution of net benefits favors lower-income individuals.

Using equation 3.6 and simplifying assumptions concerning the distribution of benefits and costs, Feldstein derived alternative forms of formula 3.5 that can be calculated from readily available data. For example, if the benefit received by an individual with income Y, $b(Y)$, is represented by

$$b(Y) = aY^\beta$$

where β is the income elasticity of demand, and if the costs of the policy are assumed to be paid out of the personal-income tax, the formula 3.5 can be written

$$B(1 + v)^{-\delta\beta} - C\pi_{\text{UDD}} \tag{3.7}$$

where B and C are the total (unweighted) dollar values of benefits and costs respectively, v is the relative variance of the income distribution, and π_{UDD} is the value of a dollar of tax revenue, measured in uniformly distributed dollars.

Once the total dollar value of a policy's benefits and costs have been calculated, formula 3.7 can easily be calculated for various values of δ. Therefore, the sensitivity of policy choices to distributional preferences can be made explicit. For example, figure 3-1 illustrates the effect of distributional preferences on the choice between two policies, F and G, where the total dollar value of net benefits is greater for F, but where a higher fraction of the net benefits of G goes to lower-income individuals. Policy F is preferred to policy G if $\delta < 0.75$ and G is preferred to F if $\delta \geq 0.75$.

The principal advantage of Feldstein's proposal is that it provides feasible procedures for aggregating over individuals using explicit distributional weights. The principal disadvantages are that the assumed distributions of benefits and costs may be only rough approximations to the true distributions, and that the formula for the marginal social utility of income omits from consideration factors such as age and race that society may consider relevant. The procedures could be elaborated to lessen the effects of these disadvantages, but only at the cost of greatly increased informational requirements.

Boadway (1976) has proposed a procedure for incorporating distributional considerations in benefit-cost analyses that does not require information on the distribution of benefits and costs for each policy. His procedure

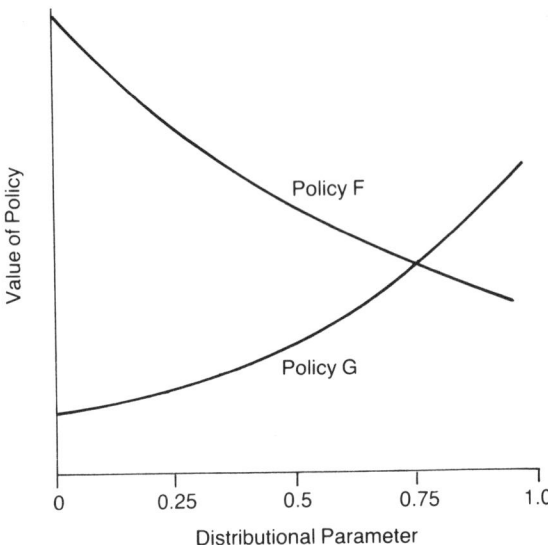

Figure 3-1. Distributional Preferences and Ranking of Policies

is based on the concept of the "distribution characteristic" of a commodity, which was originally introduced by Feldstein (1972).

The distribution characteristic of commodity k, R_k, is defined as

$$R_k = \frac{N \sum_Y x_k(Y)\alpha(Y)f(Y)}{X_k} \qquad (3.8)$$

where $x_k(Y)$ is the quantity of commodity k consumed by an individual with income Y, X_k is the total quantity consumed, and the other terms are defined as in formula 3.5. Thus the distribution characteristic of commodity k is a weighted average of marginal social utilities of income, where the weights are each individual's consumption of the commodity.[6] If the marginal social utility of income decreases as income increases, the distribution characteristic will be higher for commodities consumed by lower-income individuals than for commodities consumed by higher-income individuals.

Boadway uses the distribution characteristic to derive formulas for the effect of a policy on economic welfare that do not require information on the distribution of benefits and costs. For example, if $\partial X_k/\partial P_j = 0$ for all j not equal to k (that is, all cross-price elasticities of demand are equal to zero), and if the policy being evaluated does not alter the costs of production in the private sector, then the formula for the change in economic welfare, dW, is:

Aggregation over Individuals

$$\frac{dW}{R_n} = \sum_k \{(1 - R_k/R_n)X_k + \pi_k \partial X_k/\partial P_k\} d\pi_k + \sum_k P_k X_k^G \qquad (3.9)$$

where R_n, the distribution characteristic of commodity n, is used as a normalization factor; π_k is the tax on commodity k required to finance the policy; P_k is the price of commodity k excluding π_k (that is, the production cost of good k), and X_k^G is the public sector's production of good k.

The formulas proposed by Boadway cannot be used in their present form to evaluate air-pollution-control policies because they ignore the existence of common-property resources, such as clean air. However, the formulas could be generalized to allow their application to air-pollution-control policies, and the use of distribution characteristics may prove to be a useful procedure for incorporating distributional considerations into benefit-cost analyses.

Summary

The choice of an appropriate procedure for aggregating benefits and costs over individuals is one of the most-difficult problems in benefit-cost analysis. The simplest approach is to ignore the distributional effects of public policies and to calculate the total dollar value of net benefits without regard to whom they accrue. However, this procedure is appropriate only if the distribution of benefits does not in fact matter (that is, if the marginal social utility of income is equal for all individuals) or if costless methods of income redistribution are available. Since neither of these conditions is satisfied in practice, the distributional effects of policies should be taken into account.

At a minimum, data should be provided on the distribution of benefits and costs over relevant groups of individuals. The most promising proposals for going beyond the provision of raw data are those of Willig and Bailey, and of Feldstein. Although their procedures are not entirely satisfactory, they do provide a basis for increasing the consistency and the reviewability of the effects of distributional considerations on policy choice.

Notes

1. Freeman (1972) discusses possible procedures for determining the distribution of the benefits and costs of pollution control. Empirical studies include Dorfman and Snow (1975), Harrison and Rubinfeld (1978b), and Loehman et al. (1979). Attempts to infer the values of marginal social

utilities of income from past decisions by government include Mera (1969) and Weisbrod (1968). For a brief critique of the Mera and Weisbrod studies see Feldstein (1974a).

2. Harberger (1971) has argued that this procedure should nevertheless be used because it is consistent with national income-accounting procedures and ensures that a benefit-cost analysis carried out by one team of analysts is subject to check by others. However, the merit of consistency with national income accounting is not obvious since the purpose of benefit-cost analysis is fundamentally different; and making the analysis subject to check by others requires only that distributional considerations be made explicit, not that they be ignored. In a later article, Harberger (1978) seems to argue that distributional weights should not be used because doing so might have a large impact on the choice of policies.

3. When $h = 1$, the condition in 3.3 is equivalent to Rawls's (1971) condition that the poorest individual be made better off.

4. Willig and Bailey also derived alternative forms of the conditions in equation 3.3.

5. For easier understanding, a discrete distribution of income over individuals is assumed, rather than the continuous distribution over families assumed by Feldstein (1974a).

6. Equation 3.8 is a discrete version of Feldstein's (1972) definition of R_k. Boadway (1976) expresses R_k in terms of the marginal social utility of a specified good instead of income.

4 Aggregation of Benefits and Costs over Time

Air-pollution-control policies generally have effects that extend over a number of years. For example, a policy requiring the installation of pollution-control equipment will result in capital expenditures during the installation period, operating and maintenance costs in later years, and reductions in pollutant discharges throughout the effective life of the equipment. The actual distribution of benefits and costs over time will depend on the length of the installation period; the relative magnitues of capital, operating, and maintenance costs; the length of life of the equipment; and so forth. Therefore, decision rules are required for choosing among policies that have different distributions of benefits and costs over time.

This chapter discusses the economic theory of optimization over time and its implications for the choice of decision rules. Several decision rules frequently encountered in benefit-cost analyses are shown to be seriously defective.

Optimization over Time

The basic economic principles of optimization over time can be illustrated by the simple example of a society consisting of only one person, with only one commodity, no uncertainty, and two time periods. The results will then be generalized to cover more-realistic situations. The individual begins the present year, year 0, with a stock of the commodity; and his problem is to decide how much to consume this year and how much next year. Consumption this year will be indicated by Z_0 and consumption next year by Z_1.

The choice Z_0 and Z_1 will be determined by the individual's preferences for consumption this year and next and by the opportunity costs of consumption in each year. The opportunity cost of consuming a unit of the commodity in one year is the resulting decrease in consumption in the other year. If the portion of the commodity that is not consumed this year is simply set aside (and there are no storage costs), then the opportunity cost of one unit of Z_0 is one unit of Z_1. However, if the commodity can be invested productively (for example, if it is seed corn that can be planted), then the opportunity cost of one unit of Z_0 will be more than one unit of Z_1. Thus the opportunity cost of consumption this year will depend on the rate at which Z_0 can be transformed into Z_1 through investment.

The choices facing the individual are shown graphically in figure 4-1. Combinations of Z_0 and Z_1 that give him equal satisfaction can be indicated by *indifference curves* such as $U_0 U_0$, $U_1 U_1$, and $U_2 U_2$. Higher-numbered indifference curves correspond to greater amounts of total satisfaction (that is, utility) over the two years. The slope of an indifference curve indicates the rate at which the individual is just willing to substitute Z_0 for Z_1, and therefore is equal to (the negative of) his *marginal rate of substitution* between Z_0 and Z_1, which we will abbreviate as $\text{MRS}_{0:1}$.

If the individual consumed all of his stock of the commodity this year, he would be at point B on the horizontal axis. If he consumed none of the commodity this year, he would be at point A on the vertical axis.[1] The line between A and B shows the maximum possible combinations of consumption this year and next, and will be referred to as the *transformation curve*. The slope of the transformation curve indicates the rate at which Z_0 can be transformed into Z_1, and therefore is equal to (the negative of) the *marginal rate of transformation* between Z_0 and Z_1, which we will abbreviate as $\text{MRT}_{0:1}$. The shape of the transformation curve in figure 4-1 implies that the marginal rate of transformation decreases as the amount of consumption foregone this year increases, which is consistent with the conventional assumption that investment is subject to diminishing returns.

The Concept of Dynamic Efficiency

The greatest utility the individual can obtain is at point D, where the transformation curve is tangent to indifference curve $U_1 U_1$. The allocation of consumption between the two periods that maximizes his utility is thus Z_0^* and Z_1^*. Since he could not make himself better off by choosing any other attainable combination of Z_0 and Z_1, Z_0^* and Z_1^* represent the only efficient allocation of consumption over time. Efficiency with respect to the allocation of consumption over time will be referred to as *dynamic efficiency*.

Tangency of the indifference curve and the transformation curve implies that their slopes are equal at the point of tangency. Since the slope of the indifference is equal to $-\text{MRS}_{0:1}$ and the slope of the transformation curve is equal to $-\text{MRT}_{0:1}$, the condition for dynamic efficiency can be expressed as

$$\text{MRS}_{0:1} = \text{MRT}_{0:1} \tag{4.1}$$

Thus dynamic efficiency requires that the rate at which the individual is *willing* to substitute Z_1 for Z_0 be equal to the rate at which he is *able* to substitute Z_1 for Z_0.

Substitution of Z_1 for Z_0 involves postponing consumption by one year. Therefore, the marginal rate of substitution indicates the individual's

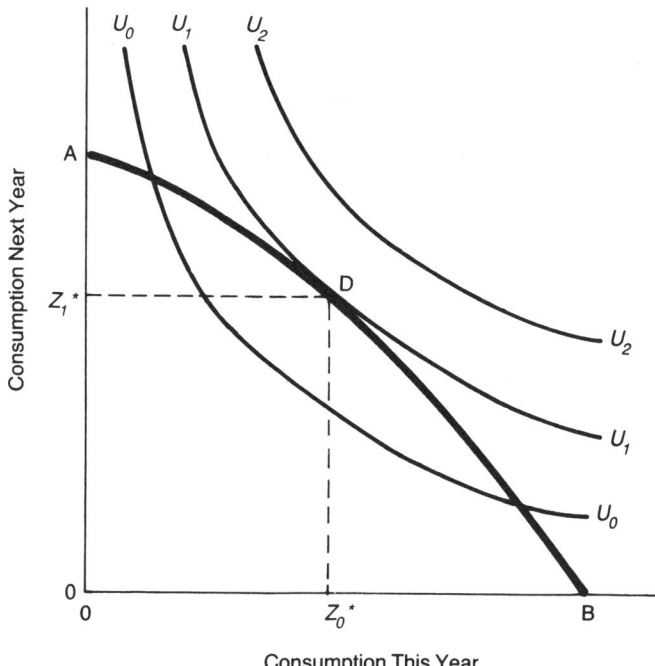

Figure 4-1. Maximization of Utility over Time

preference for consumption now rather than one year in the future. The individual's rate of time preference (RTP) is defined as

$$\text{RTP} \equiv \text{MRS}_{0:1} - 1 \qquad (4.2)$$

The rate of time preference is equal to the extra amount of Z_1 required per unit of Z_0 for the individual to be willing to postpone consumption.

Similarly, the opportunity-cost rate (OCR) is defined as

$$\text{OCR} \equiv \text{MRT}_{0:1} - 1 \qquad (4.3)$$

The opportunity-cost rate is equal to the extra amount of Z_1 that could be obtained per unit of Z_0 postponed. With these definitions the condition for dynamic efficiency can be expressed as

$$\text{RTP} = \text{OCR}$$

That is, dynamic efficiency is attained by investing up to the point where the rate of time preference is equal to the opportunity-cost rate.

The analysis of dynamic efficiency can be generalized to the case of a society consisting of many individuals by interpreting the indifference curves in figure 4-1 as society's indifference curves between Z_0 and Z_1.[2] Similarly, Z_0 and Z_1 can be interpreted as aggregates of many types of commodities in each year, and the years 0 and 1 can be interprepted as any adjacent pair of years. The condition for dynamic efficiency is then

$$\text{SMRS}_{0:1} = \text{SMRT}_{0:1} \tag{4.4}$$

where the S prefix denotes that the relevant marginal rates of substitution and transformation are those for society as a whole. Subtracting 1 from both sides of equation 4.4, the condition for dynamic efficiency can be written as

$$\text{SRTP} = \text{SOCR}$$

where SRTP is the social rate of time preference and SOCR is the social opportunity-cost rate.

In a society consisting of many individuals, consumption and investment decisions are not necessarily made by the same individuals. Therefore, dynamic efficiency is not automatically assured as it would be in a society consisting of only one individual. However, dynamic efficiency will be attained if capital markets are perfect and if there is no divergence between social and private rates of time preference and opportunity-cost rates.

If capital markets are perfect, then consumers and investors, who may or may not be the same individuals, will face the same rate of interest. Given the market rate of interest, i, a consumer will receive $1 + i$ units of Z_1 for each unit of Z_0 he foregoes. In order to maximize his total utility, he must arrange his consumption over time in such a way that the rate at which he is willing to substitute Z_1 for Z_0 (his marginal rate of substitution between Z_1 and Z_0) is equal to the rate at which he is able to substitute Z_1 for Z_0, $1 + i$. Therefore, if capital markets are perfect, and if consumers maximize their utility, then private consumption decisions will result in

$$\text{PMRS}_{0:1} = 1 + i \tag{4.5}$$

where $\text{PMRS}_{0:1}$ is the private marginal rate of substitution. Subtracting one from both sides of equation 4.5 yields the condition

$$\text{PRTP} = i \tag{4.6}$$

where PRTP is the private rate of time preference.

Similarly, if an investor wants to maximize his wealth, he will under-

Aggregation over Time

take all investments that earn him more than the rate of interest. The return to the investor is equal to his private opportunity-cost rate, POCR. Therefore, wealth-maximizing investors will invest up to the point where

$$POCR = i \tag{4.7}$$

From equations 4.6 and 4.7, perfect capital markets will result in

$$PRTP = i = POCR$$

If, in addition, private and social rates of time preference and opportunity-cost rates are equal, SRTP = PRTP and SOCR = POCR, then

$$SRTP = PRTP = i = POCR = SOCR \tag{4.8}$$

so that dynamic efficiency will be attained. The conditions necessary for dynamic efficiency may not be satisfied in most economies. However, before considering the possible reasons for failure to attain dynamic efficiency, and the implications for benefit-cost analysis, it is useful to analyze decision rules for evaluating policies when dynamic efficiency is attained.

Present-Value Criterion

Dynamic efficiency is attained when the total amounts of Z_0 and Z_1 are such that $SMRS_{0:1} = SMRT_{0:1}$. The difference between the maximum possible consumption in year 0 and the actual amount consumed can be interpreted as the total amount of investment. The problem of attaining dynamic efficiency is thus the same as that of determining the optimal amount of investment.

A given unit of investment will contribute to the attainment of efficiency if the resulting changes in Z_0 and Z_1 are such that

$$SMRT_{0:1} > SMRS_{0:1} \tag{4.9}$$

The left-hand side of inequality 4.9 can be rewritten

$$SMRT_{0:1} = \frac{\Delta Z_1}{-\Delta Z_0}$$

where ΔZ_0 is the (negative) change in consumption this year as a result of the investment and ΔZ_1 is the resulting (positive) change in consumption next year. Then inequality 4.9 can be rewritten

$$\frac{\Delta Z_1}{-\Delta Z_0} - \text{SMRS}_{0:1} > 0$$

or, multiplying by $-\Delta Z_0/\text{SMRS}_{0:1}$,

$$\Delta Z_0 + \frac{\Delta Z_1}{\text{SMRS}_{0:1}} > 0 \tag{4.10}$$

Recalling that $\text{SRTP} = \text{SMRS}_{0:1} - 1$, inequality 4.10 can be rewritten

$$\Delta Z_0 + \frac{\Delta Z_1}{1 + \text{SRTP}} > 0 \tag{4.11}$$

The division of ΔZ_1 by $1 + \text{SRTP}$ reflects the premium that society requires in order to be willing to postpone consumption for one year. Thus dividing by $1 + \text{SRTP}$ can be interpreted as discounting ΔZ_1 because it occurs in the future, where the rate of discount is equal to the social rate of time preference.

The notation can be simplified by defining $\gamma \equiv \text{STRP}$. Then inequality 4.11 is

$$\Delta Z_0 + \frac{1}{1 + \gamma} \Delta Z_1 > 0 \tag{4.12}$$

The term $(1/1 + \gamma)\Delta Z_1$ in inequality 4.12 can be interpreted as the value in the present (year 0) of a change in consumption one year in the future. Let $V_0(\Delta Z_t)$ represent the present value, in year 0, of a change in consumption in year t. Then

$$V_0(\Delta Z_1) = \frac{1}{1 + \gamma} \Delta Z_1 \tag{4.13}$$

The present value of a change in consumption in year 0 is simply ΔZ_0, since no discounting for time preference is required;

$$V_0(\Delta Z_0) = \Delta Z_0 \tag{4.14}$$

Using definitions 4.13 and 4.14 permits the left-hand side of inequality 4.12 to be rewritten

$$V_0(\Delta Z_0) + V_0(\Delta Z_1) = PV \tag{4.15}$$

where PV is the total present value of the changes in consumption resulting

Aggregation over Time

from the investment. Therefore, the rule to be used in evaluating an investment can be restated as follows:

Calculate the total present value of the changes in present and future consumption using the social rate of time preference as the discount rate. The investment is worthwhile if its total present value is greater than zero.

This rule can be used to evaluate investments resulting in changes in consumption over more than two years. The general form of 4.15 can be written

$$PV = V_0(\Delta Z_0) + V_0(\Delta Z_1) + V_0(\Delta Z_2) + \ldots + V_0(\Delta Z_T) \quad (4.16)$$

where T is the last year in which a change in consumption occurs as a result of the investment. Generalizing equation 4.13, for any two adjacent years j and k, the value in year j of a change in consumption in year k is

$$V_j(\Delta Z_k) = \frac{1}{1+\gamma} \Delta Z_k \quad (4.17)$$

For example, the value in year 1 of a change in consumption in year 2 is

$$V_1(\Delta Z_2) = \frac{1}{1+\gamma} \Delta Z_2$$

Applying 4.17 sequentially, the value in year 0 of a change in consumption in year 2 is

$$V_0(\Delta Z_2) = V_0[V_1(\Delta Z_2)] = V_0\left[\frac{1}{1+\gamma} \Delta Z_2\right]$$
$$= \frac{1}{1+\gamma} \frac{1}{1+\gamma} \Delta Z_2 = \frac{1}{(1+\gamma)^2} \Delta Z_2 \quad (4.18)$$

and the general formula for the present value in year 0 of a change in consumption in year t is

$$V_0(\Delta Z_t) = \frac{1}{(1+\gamma)^t} \Delta Z_t \quad (4.19)$$

Substituting in equation 4.16 using equation 4.19, the present-value of a policy that affects consumption over a period of T years is

$$PV = \frac{1}{(1+\gamma)^0} \Delta Z_0 + \frac{1}{(1+\gamma)^1} \Delta Z_1 + \ldots + \frac{1}{(1+\gamma)^T} \Delta Z_T$$

or, more compactly

$$PV = \sum_{t=0}^{T} \frac{1}{(1+\gamma)^t} \Delta Z_t \qquad (4.20)$$

As shown in equations 4.8, if dynamic efficiency is attained, the social rate of time preference, γ, will be equal to the market rate of interest, i, and equation 4.20 can be rewritten as

$$PV = \sum_{t=0}^{T} \frac{1}{(1+i)^t} \Delta Z_t \qquad (4.21)$$

In practice, data on the change in consumption in year t, ΔZ_t, will not be directly available. Instead, data will be available on the net benefits of a policy, $B_t - C_t$, where B_t and C_t are total benefits and total costs, respectively, in year t. In general, $B_t - C_t$ will not be equal to ΔZ_t because some part of the net benefits of a policy may go to investment rather than consumption. However, when the conditions for dynamic efficiency are satisfied, the social value of one dollar's worth of investment is equal to the social value of one dollar's worth of consumption, and $B_t - C_t$ can be used in place of ΔZ_t in equality 4.21.[3] The right-hand side of equation 4.21 is then

$$\sum_{t=0}^{T} \frac{1}{(1+i)^t} (B_t - C_t) \qquad (4.22)$$

In calculating present values, it is necessary to adjust for the effects of inflation on the dollar values of benefit and costs. As a result of inflation, the current-dollar value of benefits and costs will increase over time even if benefits and costs measured in constant dollars remain unchanged. For example, if the annual rate of inflation is equal to f, then the current-dollar value of net benefits in year t will be $(1 + f)^t(B_t - C_t)$, where $(B_t - C_t)$ represents net benefits measured in constant dollars.

Similarly, inflation will cause the nominal rate of interest to be greater than the real rate of interest, because part of the nominal return to savers is negated by the increase over time in the price of commodities. The relation between the nominal and real rates of interest is given by

$$i = \frac{1+g}{1+f} - 1 \qquad (4.23)$$

where i is the real rate of interest and g is the nominal rate.[4]

If net benefits are measured in constant dollars, as is customary, then

Aggregation over Time

the real rate of interest should be used in calculating net present values. If net benefits are measured in current dollars, discounting by the nominal rate of interest will yield results identical to those obtained using constant-dollar net benefits and the real rate of interest, because

$$\sum_{t=0}^{T} \frac{1}{(1+g)^t} (1+f)^t (B_t - C_t) = \sum_{t=0}^{T} \frac{1}{[(1+i)(1+f)]^t} (1+f)^t (B_t - C_t)$$

$$= \sum_{t=0}^{T} \frac{1}{(1+i)^t} (B_t - C_t)$$

However, use of a nominal rate of interest to discount net benefits expressed in constant dollars, as has occurred surprisingly often in practice, is incorrect and can lead to serious underestimates of the net present value of a policy.

Net-Present-Value Decision Rules

Formula 4.22 is referred to as the *net present value* of the investment. The net-present-value decision rule for evaluating an investment when the conditions for dynamic efficiency are satisfied in the rest of the economy can be stated as

Rule I.A. *An investment is worthwhile if its net present value is greater than zero.*

In choosing between mutually exclusive investments (for example, two alternative investments in air-pollution-control equipment for an industrial facility), the net-present-value decision rule is;

Rule I.B. *Choose the investment with the largest net present value.*[5]

Application of the net-present-value decision rules is illustrated in table 4-1 for two hypothetical investments, X and Y. Benefits and costs are assumed to be measured in constant dollars. The net present value of each investment is calculated using a real interest rate of 5 percent. (Determination of the appropriate value of the rate of interest to use in practice will be discussed later on.) As shown in table 4-1, the net present value is $157 for investment X and $446 for investment Y. By decision-rule I.A, both investments are worthwhile. By decision-rule I.B, if the investments are mutually exclusive, investment Y should be chosen.

Table 4-1
Calculation of Net Present Value

	Year 0	Year 1	Year 2	Year 3	Year 4 Onwards
I. Investment X					
B_t	0	475	475	475	0
C_t	1,000	50	50	50	0
$B_t - C_t$	−1,000	425	425	425	0
II. Investment Y					
B_t	0	2,250	2,250	2,250	0
C_t	5,000	250	250	250	0
$B_t - C_t$	−5,000	2,000	2,000	2,000	0

Net Present Value ($i = 0.05$):

Investment X

$$\sum_{t=0}^{T} \frac{1}{(1+i)^t}(B_t - C_t) = -1,000 + \left(\frac{1}{1.05}\right)425 + \left(\frac{1}{1.05^2}\right)425 + \left(\frac{1}{1.05^3}\right)425 = 157$$

Investment Y

$$\sum_{t=0}^{T} \frac{1}{(1+i)^t}(B_t - C_t) = -5,000 + \left(\frac{1}{1.05}\right)2,000 + \left(\frac{1}{1.05^2}\right)2,000 + \left(\frac{1}{1.05^3}\right)2,000 = 446$$

Alternative Investment-Decision Rules

The investment-decision rules based on the concept of net present value are consistent with the maximization of well-being over time and provide a theoretically correct basis for the evaluation of investments. Although the superiority of the net-present-value rules is now generally recognized by economists, alternative sets of decision rules are sometimes used in evaluating air-pollution-control policies. As will be shown, these alternative sets of rules can result in serious errors in policy analysis.

Benefit-Cost-Ratio Decision Rules

One of these alternative sets of investment-decision rules is based on the concept of the benefit-cost ratio. The benefit-cost ratio is equal to the present value of benefits, B, divided by the present value of costs, C, where

$$B = \sum_{t=0}^{T} \frac{1}{(1+i)^t} B_t$$

and

$$C = \sum_{t=0}^{T} \frac{1}{(1+i)^t} C_t$$

The benefit-cost-ratio decision rules are:

Rule II.A. *An investment is worthwhile if its benefit-cost ratio is greater than one.*

and

Rule II.B. *In choosing among mutually exclusive investments, choose the one with the largest benefit-cost ratio.*

Decision-rule II.A is mathematically equivalent to rule I.A, because $B/C > 1$ implies $B - C > 0$. Therefore, use of rule II.A will correctly indicate whether or not an investment is worthwhile. However, the benefit-cost-ratio rule for choosing between mutually exclusive investments, rule II.B, is not equivalent to the net-present-value rule, I.B, and its use can result in incorrect choices among investments.

Because the benefit-cost ratio does not reflect the scale of an investment, the use of rule II.B may lead to an incorrect choice when investments of different scale are being compared. This can be illustrated with the data for investments X and Y from table 4-1. As shown in part A of table 4-2, the benefit-cost ratio is 1.14 for investment X and 1.08 for investment Y. Therefore, use of rule II.B would result in the choice of investment X even though investment Y has a larger net present value. The inconsistency occurs because the benefit-cost ratio does not reflect the larger scale of investment Y, which more than offsets the lower ratio of benefits to costs.[6]

The sensitivity of the benefit-cost ratio to arbitrary decisions concerning the classification of benefits and costs makes the use of rule II.B inadvisable even when the scale of the investments being compared is the same. To illustrate, suppose that a proposed air-pollution-control policy would increase water pollution as a result of the discharge into the water of pollutants that otherwise would have been discharged into the air. The increase in water pollution could equally well be classified either as an addition to the costs of the policy or as a subtraction from its benefits—that is, either as a positive cost or as a negative benefit. If it is classified as a negative benefit, then the benefit-cost ratio will be $(B - D)/C$, where D is the present value of the water pollution. If it is classified as a positive cost, then the benefit-cost ratio will be $B/(C + D)$.

In general, the value of these two ratios will not be equal.[7] If this policy were being compared with a mutually exclusive policy for which the benefit-

Table 4-2
Alternative Investment-Decision Rules

	Investment X^a	Investment Y^a
A. Benefit-cost ratio ($i = 0.05$):		
$B = \sum_{t=0}^{T} \frac{1}{(1+i)^t} B_t$	1,294	6,127
$C = \sum_{t=0}^{T} \frac{1}{(1+i)^t} C_t$	1,136	5,681
	$B/C = 1.14$	$B/C = 1.08$
B. Internal rate of return, r:		
$\sum_{t=0}^{T} \frac{1}{(1+r)^t}(B_t - C_t) = 0$	$r = 0.132$	$r = 0.097$
C. Payback period, t^*:		
$\sum_{t=0}^{t^*} B_t - C_t = 0$	$t^* = 2.35$	$t^* = 2.50$

[a]See table 4-1.

cost ratio lay between the values of $(B - D)/C$ and $B/(C + D)$, the choice between the policies could be determined by a purely arbitrary decision with respect to the classification of the water pollution as a negative benefit or a positive cost.

Internal-Rate-of-Return Decision Rules

Another set of decision rules that is sometimes used to evaluate investments is based on the concept of the internal rate of return. An investment's internal rate of return is defined to be the rate of discount that makes the net present value of the investment equal to zero. Thus the internal rate of return, r, is calculated by solving

$$\sum_{t=0}^{T} \frac{1}{(1+r)^t}(B_t - C_t) = 0 \qquad (4.24)$$

for r. The decision rules based on the internal rate of return are

Aggregation over Time

Rule III.A. *An investment is worthwhile if its internal rate of return, r, is greater than the market rate of interest, i.*

and

Rule III.B. *In choosing among mutually exclusive investments, choose the one with the largest internal rate of return.*

If an investment has a "conventional" distribution of net benefits over time, rule III.A will be mathematically equivalent to rule I.A, because $r > i$ will imply $\sum_{t=0}^{T}((1/1 + i)^t)(B_t - C_t) > 0$. The distribution of net benefits over time is "conventional" if net benefits are at first negative, then turn positive, and do not change sign again subsequently. Investments X and Y in table 4-1 both have conventional distributions of net benefits over time, and each has a positive net present value when the rate of discount is less than the internal rate of return.

However, if the distribution of net benefits over time is not conventional, rules I.A and III.A will not be equivalent; in that case, the use of rule III.A to evaluate an investment may lead to an incorrect decision. To illustrate, suppose that a pesticide-spraying program reduces crop damage in the short run but has adverse effects on human health in the long run. To keep the example simple, let the net benefits of the policy be $1,000 in year 0, −$1,100 in year 1, and 0 in subsequent years. The internal rate of return of this investment would be 0.10, and rule III.A would indicate that it was worthwhile if $i < 0.10$. However, if $i < 0.10$, the investment would have a negative net present value and should not be undertaken.

A second type of investment with a nonconventional distribution of net benefits over time is one for which net benefits begin negative, turn positive, and then turn negative again. This might arise as a result of environmental regulations that required restoration of damage caused during the life of a project (for example, reclamation of land used for strip mining). For example, suppose that the net benefits of an investment are −$1,000 in year 0, $2,900 in year 1, −$2,000 in year 2, and 0 in subsequent years. From equation 4.24, the equation to be solved for the internal rate of return is

$$-1{,}000 + \frac{2{,}900}{1 + r} - \frac{2{,}000}{(1 + r)^2} = 0 \qquad (4.25)$$

Equation 4.25 has two roots, $r_1 = 0.13$ and $r_2 = 0.77$.[8] If the market rate of discount, i, were less than 0.13, rule III.A would indicate that the investment was worthwhile, since $r_1 > i$ and $r_2 > i$. However, the net present value would be positive only if $0.13 < i < 0.77$.

As these examples demonstrate, use of rule III.A to evaluate an investment is appropriate only if the distribution of net benefits over time is conventional. Even if this condition is satisfied, use of rule III.B to evaluate mutually exclusive investments can result in incorrect investment choices.

Because the internal rate of return does not reflect the scale of an investment, the use of rule III.B to choose among investments of different scale may lead to incorrect choices. This can be seen from the data for investments X and Y of table 4-1. As shown in part B of table 4-2, the internal rate of return is 13.2 percent for investment X and 9.7 percent for investment Y. Therefore, use of rule III.B would result in the choice of investment X, even though investment Y has a larger net present value.[9]

Since the internal rate of return does not reflect the prevailing market rate of interest, the use of rule III.B to evaluate mutually exclusive investments may lead to incorrect choices even if the investments do not differ in scale. This can be illustrated using the data in table 4-3 for two hypothetical investments, F and G, that have identical costs but different distributions of net benefits over time. The internal rate of return is 24.5 percent for investment F and 38.6 percent for investment G. Therefore, use of rule III.B would result in the choice of investment G. However, as shown in figure 4-2, the net present value of investment F is greater than the net present value of investment G for all interest rates less than 11.0 percent. Therefore, investment G is the correct choice only if i is greater than 11.0 percent.

Payback-Period Decision Rules

The final set of potentially misleading decision rules that will be discussed is based on the concept of the payback period. The payback period is defined to be the number of years required for the sum of the undiscounted benefits

Table 4-3
Internal Rates of Return

	Year 0	Year 1	Year 2	Year 3 Onwards	Internal Rate of Return
Investment F					$r = 0.245$
Benefits	0	0	1,550	0	
Costs	1,000	0	0	0	
Net benefits	−1,000	0	1,550	0	
Investment G					$r = 0.386$
Benefits	0	1,350	50	0	
Costs	1,000	0	0	0	
Net benefits	−1,000	1,350	50	0	

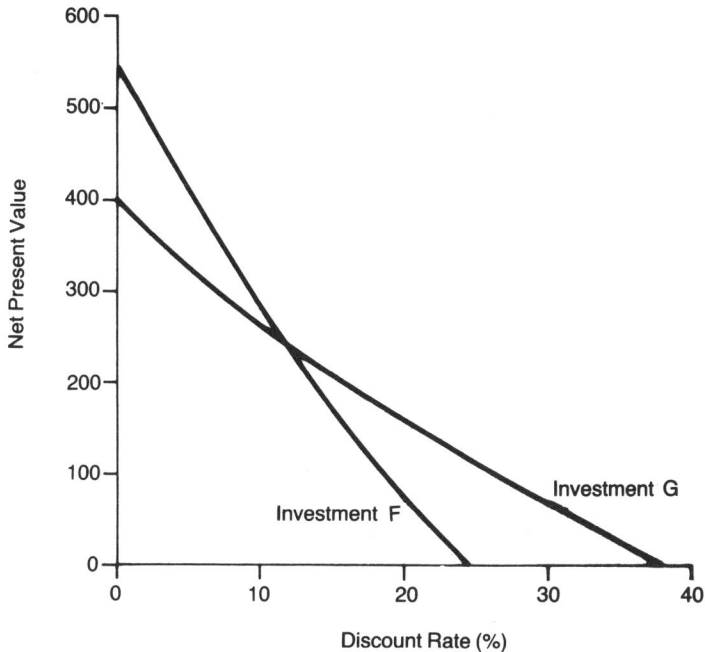

Figure 4-2. Discount Rates and Ranking of Investments

to be equal to the sum of the undiscounted costs. The calculation of the payback periods for investments X and Y is shown in table 4-2. The decision rules based on the payback period are:

Rule IV.A. *An investment is worthwhile if the payback period is less than some (arbitrary) critical value.*

and

Rule IV.B. *In choosing among mutually exclusive projects, choose the one with the smallest payback period.*

Because the payback-period decision rules consider only benefits and costs that occur prior to payback and do not reflect the rate of interest, they provide a very poor basis for evaluating investments. For example, suppose a policy resulted in costs of $1,000 in year 0, benefits of $1,000 in year 1, and 0 benefits and costs in subsequent years. The payback period would be only one year, and the policy would be determined to be worthwhile using rule IV.A if the critical value for the payback period were greater than 1, as is usually the case. However, the policy would have a negative net present value at any positive rate of interest, and therefore should not be undertaken.

If rule IV.B is used to choose among mutually exclusive investments, an additional source of error is that the payback period does not reflect differences in the scale of the investments being compared. Even if the investments being compared had the same scale, the failure of the payback period to reflect the full stream of net benefits makes it an inappropriate basis for comparing policies.

Comparison of Decision Rules

The set of decision rules based on net present value provides a conceptually correct basis for the evaluation of policies that have effects on consumption in more than one year. Alternative sets of decision rules are sometimes used in practice, but are inferior to the net-present-value rules.

Decision rules based on the benefit-cost ratio can be used to determine whether an investment is worthwhile but may result in errors in choosing among mutually exclusive investments because the value of the benefit-cost ratio reflects arbitrary decisions with respect to the classification of benefits and costs while failing to reflect differences in scale. Decision rules based on the internal rate of return can be used to determine whether an investment is worthwhile, provided that the distribution of net benefits over time is conventional. However, use of the internal rate of return to choose among mutually exclusive investments can result in errors because the internal rate of return reflects neither the scale of an investment nor the prevailing market rate of interest. Decision rules based on the payback period can result in errors both in determining if an investment is worthwhile and in choosing among mutually exclusive investments, because the payback period ignores all benefits and costs occurring after payback, and reflects neither the scale of an investment nor the rate of interest.

Up to now, the discussion has been based on the assumption that the conditions for dynamic efficiency are satisfied. The next section discusses the reasons for expecting that dynamic efficiency will not be attained in actual economies, and the following section examines the necessary modifications to the net-present-value decision rules when dynamic efficiency is not attained.

Sources of Dynamic Inefficiency

The necessary and sufficient condition for dynamic efficiency is that the social rate of time preference be equal to the social opportunity-cost rate.

Aggregation over Time

This condition will be satisfied if capital markets are perfect and if social and private rates of time preference and opportunity cost rates are equal, because then

$$\text{SRTP} = \text{PRTP} = i = \text{POCR} = \text{SOCR} \qquad (4.26)$$

If one or more of the equalities in 4.26 does not hold, dynamic efficiency will not be attained.

As discussed earlier in this chapter, perfect capital markets would result in the private rate of time preference and private opportunity-cost rate being equal to the (same) rate of interest, PRTP = i = POCR. Actual capital markets are unlikely to result in these equalities being satisfied. Government regulations on the interest rates banks can pay and the types of loans they can make interfere with the free flow of funds. Also, some financial institutions may be large enough relative to their markets to be able to insert a monopoly markup between the rate of interest paid to savers, i_S, and the rate charged to firms borrowing for investment, i_B, so that

$$\text{PRTP} = i_S < i_B = \text{POCR}$$

Taxes on income from capital will prevent the equality POCR = SOCR from being satisfied. The private opportunity-cost rate is equal to the after-tax rate of return, whereas the social opportunity-cost rate is equal to the before-tax rate of return. Therefore,

$$\text{POCR} = (1 - \pi)\,\text{SOCR} < \text{SOCR}$$

where π is the effective marginal tax rate on capital income.

The relationship between the social rate of time preference and the private rate of time preference is more subjective but also may be a source of dynamic inefficiency. The social rate of time preference is the rate *society should use* in comparing the value of consumption in different years, and the private rate of time preference is the rate that *individuals actually use*. Therefore, equality of the social and private rates of time preference is equivalent to a policy decision to accept the private rates as reflecting society's preferences.

Several arguments have been advanced in favor of rejecting the private rate of time preference as the measure of society's preferences. One argument is that the private rate is too high because it reflects individuals' risks of dying. Since society does not die, it is argued, the rate of time preference used for social decisions should be lower than the private rate. However, this argument is inconsistent with the assumption, discussed in chapter 2, that social welfare is a function of the well-being of the individuals making up the society.

A second line of argument is that the private rate of time preference is too high because it does not reflect adequate concern for the well-being of future generations. One rationale advanced for this position is that the government should act as a trustee for future generations and should ensure that the rate of time preference reflects their interests as well as those of the present generation. An alternative rationale has been advanced by Marglin (1963a), based only on the current generation's preferences. He argues that the private rate of time preference does not reflect these preferences correctly because provision for future generations by one individual provides positive externalities for other members of the present generation. Because markets tend to provide less than the optimal quantity of a good that involves positive externalities, there will be too little provision for future generations; therefore, the government should use a social rate of time preference less than the private rate.[10]

Not all economists agree with Marglin's conclusion that the social rate of time preference is less than the private rate of time preference. There also is controversy concerning the magnitude of the effects of capital-market imperfections on the relationship between the private rate of time preference and the private opportunity-cost rate. However, almost all economists agree that the difference between the private opportunity-cost rate and the social opportunity-cost rate created by taxes on capital income is an important source of dynamic inefficiency. Therefore, whether or not they accept the importance of all the possible sources of inefficiency previously discussed, there is a general consensus among economists that the social rate of time preference is less than the social opportunity-cost rate,

$$\text{SRTP} < \text{SOCR} \qquad (4.27)$$

Inequality 4.27 implies that the actual level of investment is less than the level corresponding to dynamic efficiency. Policies that would reduce the extent of dynamic inefficiency include the elimination of taxes on capital income and the deregulation of interest rates.[11] Analysis of whether such policies would increase social welfare is beyond the scope of this book. Instead, our concern is with the implications of the existence of dynamic inefficiency for the evaluation of investments in air-pollution control.

Implications of Dynamic Inefficiency

The appropriate measure of the social value of an investment is the present value of its effects on consumption, equation 4.20,

Aggregation over Time

$$\sum_{t=0}^{T} \frac{1}{(1+\gamma)^t} \Delta Z_t \qquad (4.28)$$

where γ is the social rate of time preference and ΔZ_t is the effect of the investment on consumption in year t.[12] If the conditions for dynamic efficiency are satisfied, an equivalent measure is provided by the net-present-value formula, 4.22:

$$\sum_{t=0}^{T} \frac{1}{(1+i)^t} (B_t - C_t) \qquad (4.29)$$

where i is the market rate of interest and $B_t - C_t$ is the dollar value of net benefits in year t. However, if dynamic efficiency is not attained, formula 4.29 will not be equivalent to 4.28 and its use may result in incorrect investment decisions.

There are two possible sources of error in using formula 4.29 when dynamic efficiency is not attained. First, the social rate of time preference may not be equal to the market rate of interest. If it is not, γ should be substituted for i in formula 4.29. Although it is conceptually simple, this requires determination of the appropriate value for γ. The problem of choosing a value for γ will be discussed later in this chapter.

Second, dynamic inefficiency implies that the social value of one dollar's worth of private investment will not be equal to the social value of one dollar's worth of consumption. Therefore, the effects of a public investment on consumption and private investment in each year must be identified and expressed in terms of their corresponding social values.

The social value in any year, τ, of one dollar's worth of consumption in that year is simply one dollar. The social value in year τ of a dollar's worth of private investment in that year is the present value in year τ of the investment's effects on future consumption, calculated using formula 4.28.

The effects of one dollar's worth of private investment on future consumption will depend on the amount of income produced by the investment each year and the fraction of that income reinvested rather than consumed. The rate at which a private investment can produce income is equal, by definition, to the social opportunity-cost rate. Therefore, an increase of one dollar in private investment in year τ will result in an increase in income in year $\tau + 1$ of ρ dollars, where ρ is the social opportunity-cost rate. Of this amount, a fraction s will be invested and a fraction $1 - s$ will be consumed. Therefore, as shown in table 4-4, there will be a further increase in private investment in year $\tau + 1$ of $s\rho$ dollars (so the total increase by the end of year $\tau + 1$ will be $1 + s\rho$ dollars) and an increase in consumption of $(1 - s)\rho$ dollars. In the subsequent year, $\tau + 2$, income will be increased by $\rho(1 +$

Table 4-4
Calculation of the Social Value of Private Investment

	Year τ	Year $\tau+1$	Year $\tau+2$	Year $\tau+m$
Increase in income	0	ρ	$\rho(1+s\rho)$	$\rho(1+s\rho)^{m-1}$
Increase in investment	1	$s\rho$	$s\rho(1+s\rho)$	$s\rho(1+s\rho)^{m-1}$
Increase in consumption ($\Delta Z_{\tau+t}$)	0	$(1-s)\rho$	$(1-s)\rho(1-s\rho)$	$(1-s)\rho(1+s\rho)^{m-1}$

$$V_\tau = \sum_{t=0}^{\infty} \frac{1}{(1+\gamma)^t} \Delta Z_{\tau+t}$$

$$= 0 + \frac{(1-s)\rho}{(1+\gamma)} + \frac{(1-s)\rho(1+s\rho)}{(1+\gamma)^2} + \ldots + \frac{(1-s)\rho(1+s\rho)^{m-1}}{(1+\gamma)^m} + \ldots$$

$$= \frac{(1-s)\rho}{1+\gamma} \sum_{k=0}^{\infty} \left(\frac{1+s\rho}{1+\gamma}\right)^k$$

where ρ = Social opportunity-cost rate.
s = Marginal propensity to save.
γ = Social rate of time preference.

$s\rho$) dollars, of which s dollars will be invested, $(1-s)$ dollars will be consumed, and so on.[13]

As shown in table 4-4, the formula for the social value of one dollar's worth of private investment, in the year in which it occurs, is

$$V = \frac{(1-s)\rho}{1+\gamma} \sum_{k=0}^{\infty} \left(\frac{1+s\rho}{1+\gamma}\right)^k \quad (4.30)$$

If $s\rho$ is less than γ, equation 4.30 will converge to

$$V = \frac{(1-s)\rho}{\gamma - s\rho} \quad (4.31)$$

If dynamic efficiency were attained, the social rate of time preference, γ, would be equal to the social opportunity-cost rate, ρ. Therefore, from equation 4.31, the social value of one dollar's worth of private investment would be one dollar. However, when dynamic efficiency has not been

Aggregation over Time

attained because γ is less than ρ, this implies that there is too little total investment; and the social value of one dollar's worth of private investment will be greater than one. For example, if $\gamma = 0.02$, $\rho = 0.08$, and $s = 0.10$, then the social value of one dollar's worth of private investment would be equal to $6.00.

Therefore, when dynamic efficiency is not attained, the social values of the benefits and costs of a public policy will depend on their effects on consumption and private investment. If θ is the fraction of each dollar of costs coming from private investment and $1 - \theta$ is the fraction coming from consumption, then the social value of one dollar of costs in the year in which they are incurred will be

$$V_C = \theta V + (1 - \theta)1 = 1 + \theta(V - 1) \tag{4.32}$$

Similarly, if ϕ is the fraction of each dollar of benefits going to private investment and $1 - \phi$ is the fraction going to consumption, then the social value of one dollar of benefits will be

$$V_B = \phi V + (1 - \phi)1 = 1 + \phi(V - 1) \tag{4.33}$$

The social value of the net benefits of a policy in year τ will then be equal to

$$B_\tau V_B - C_\tau V_C \tag{4.34}$$

In summary, when dynamic efficiency is not attained, use of the present-value formula, 4.29, to evaluate investments may result in incorrect decisions because the social rate of time preference, γ, may not be equal to the market rate of interest, i, and because the social value of one dollar's worth of private investment will not be equal to the social value of one dollar's worth of consumption. The first source of error can be eliminated by substituting γ for i in formula 4.29, and the second can be eliminated by substituting the social value of net benefits, formula 4.34, for net benefits. The formula to be used in evaluating investments is then

$$\sum_{t=0}^{T} \frac{1}{(1 + \gamma)^t} (B_t V_B - C_t V_C) \tag{4.35}$$

Net-Social-Benefit Criterion

Formula 4.35 will be referred to as the *net social benefit* of an investment. Because benefits and costs are expressed in terms of units of consumption using V_B and V_C, the net-social-benefit formula, 4.35, is equivalent to the theoretically correct measure of the social value of an investment, formula 4.28. Therefore, the net-social-benefit formula provides an appropriate basis for evaluating investments when dynamic efficiency is not attained.

Unfortunately, application of formula 4.35 requires information that is difficult to obtain. Determination of the value of θ, the fraction of costs coming from private investment, requires knowledge both of the methods used to finance the investment and of the ultimate effect of each method of finance on private investment. Similarly, determination of the value of ϕ requires knowledge of the impact of each type of benefit on investment and consumption. Determination of the value of V, the social value of one dollar's worth of private investment, requires knowledge of the social rate of time preference, the social opportunity-cost rate, and the marginal propensity to save.

Given the difficulties involved in calculating 4.35, it is useful to consider the probable magnitude of the errors that would occur if a simpler formula were used. For example, if no attempt were made to adjust benefits and costs for the social values of consumption and private investment, formula 4.35 would become

$$\sum_{t=0}^{T} \frac{1}{(1 + \gamma)^t} (B_t - C_t) \qquad (4.36)$$

This is simply the net-present-value formula, 4.29, with the social rate of time preference used as the discount rate.

If the conditions for dynamic efficiency were satisfied, formula 4.35 would reduce to 4.36 because $\gamma = \rho$ implies $V = 1$, and therefore $V_B = V_C = 1$. Also, since γ would equal i, formula 4.36 would be identical to 4.29. These results reaffirm our earlier conclusion that use of the net-present-value formula results in correct investment decisions when dynamic efficiency is attained.

In considering the magnitude of errors that might occur using formula 4.36 when dynamic efficiency is not attained, it is convenient to rewrite 4.35 as

$$V_B \sum_{t=0}^{T} \frac{1}{(1 + \gamma)^t} (B_t - C_t \frac{V_C}{V_B}) \qquad (4.37)$$

The multiplicative factor, V_B, will not affect either the sign of the net social benefit or the rank ordering of investments that would be obtained using it. Therefore, the only effective difference between formulas 4.37 and 4.36 is

Aggregation over Time

the multiplication of costs by the factor V_C/V_B in 4.37 and not in 4.36. The magnitude of the errors that might occur using formula 4.36 depends solely on the ratio of V_C to V_B.

If the effect on private investment of one dollar of costs were equal to the effect of one dollar of benefits (that is, if $\theta = \phi$), then, from equations 4.32 and 4.33, V_C would equal V_B. Therefore, V_C/V_B would equal 1, and use of the net-present-value formula, calculated using the social rate of time preference as the rate of discount, would yield correct investment decisions whether or not dynamic efficiency were attained.[14]

Therefore, when $\theta = \phi$, the effects of a public policy on private investment can be ignored even if the social value of private investment is much higher than the social value of consumption. The reason is that the negative effects on private investment of the costs of the public policy are exactly offset by the positive effects of the benefits of the policy. However, in the more general case of $\theta \neq \phi$, ignoring these effects by using formula 4.36 in place of 4.35 may result in incorrect investment decisions.

When $\theta \neq \phi$, the size of the ratio V_C/V_B will depend on both the difference between θ and ϕ and the social value of one dollar of private investment, V. Bradford (1975) concluded that the value of V was unlikely to be larger than $1.10, and therefore that the errors involved in using formula 4.36 would be small even if θ and ϕ took on extreme values (for example, $\theta = 1.0, \phi = 0$). However, as pointed out by Mendelsohn (1981), Bradford made a critical error in calculating V. Using the correct formula, 4.31, much larger values for V are obtained for plausible values of γ, ρ, and s.

As noted previously, if $\gamma = 0.02$, $\rho = 0.08$, and $s = 0.10$, then the value of V is $6.00. The value of the ratio V_C/V_B for alternative combinations of θ and ϕ when $V = \$6.00$ are shown in table 4-5. As indicated by the values of V_C/V_B in this table, use of formula 4.36 to evaluate investments may result in serious errors. For example, if $V = \$6.00$, $\theta = 0.6$, and $\phi = 0.2$, then $V_C/V_B = 2.0$ and use of formula 4.36 would understate the social value of costs relative to benefits by one-half, leading to the acceptance of some policies with negative net social benefits. If $\theta = 0.2$ and $\phi = 0.6$, then $V_C/V_B = 0.5$ and use of formula 4.36 would overstate the social value of costs by a factor of two, leading to the rejection of some policies with positive net social benefits.

It is important to note that the possible errors involved in the use of formula 4.36 could not be eliminated by the use of a different discount rate. Therefore, as Feldstein (1974b) pointed out, the frequent suggestion that policies be evaluated by calculating their net present values using a weighted average of γ and ρ as the discount rate would not avoid errors in investment decisions arising from differences in the social values of consumption and investment.

Table 4-5
Ratio of V_C to V_B When $V = \$6.00$

	Value of θ					
Value of ϕ	0.0	0.2	0.4	0.6	0.8	1.0
0	1.00	2.00	3.00	4.00	5.00	6.00
0.2	0.50	1.00	1.50	2.00	2.50	3.00
0.4	0.33	0.67	1.00	1.33	1.67	2.00
0.6	0.25	0.50	0.75	1.00	1.25	1.50
0.8	0.20	0.40	0.60	0.80	1.00	1.20
1.0	0.17	0.33	0.50	0.67	0.83	1.00

If the necessary information on θ, ϕ, and V is available, then the net-social-benefit formula, 4.35, can be used to obtain correct investment decisions. An investment is worthwhile if its net social benefit is greater than zero; and in choosing among mutually exclusive investments, the one with the greatest net social benefit should be chosen. If the values of θ, ϕ, and V are not known, investments can be evaluated by calculating formula 4.35 under the assumption that $V_C/V_B = 1$ and then testing the sensitivity of the resulting investment decisions to alternative plausible values for V_C/V_B.

In either case, calculation of formula 4.35 requires information on the social rate of time preference, γ. Although the social rate of time preference can not be directly observed, the rate of interest earned on savings, i_D, provides an estimate of its upper bound since

$$\text{SRTP} \leq \text{PRTP} = i_D$$

An appropriate measure of i_D in the United States is provided by the real rate of interest on six-month Treasury bills, which has been estimated to be approximately 2.0 percent (Fama 1976; Carlson 1977).

Summary

Dynamic efficiency is attained only if the social rate of time preference equals the social opportunity-cost rate. If dynamic efficiency is attained, the net-present-value formula provides an appropriate basis for aggregation of benefits and costs over time. An investment is worthwhile if its net present value is greater than zero; and in choosing among mutually exclusive investments, the one with the largest net present value should be chosen.

Aggregation over Time

Alternative bases for investment decision rules are the benefit-cost ratio, the internal rate of return, and the payback period. Although frequently encountered in practice, the alternative decision rules are seriously defective, especially for use in choosing among mutually exclusive investments.

Dynamic efficiency may fail to be attained because provision for future generations involves positive externalities, because capital markets are imperfect, or because of taxes on income from capital. Although economists differ on the importance of each of these possible sources of dynamic inefficiency, there is general agreement that dynamic efficiency is not attained and that the social rate of time preference is less than the social opportunity-cost rate.

When the social rate of time preference is less than the social opportunity-cost rate, one dollar's worth of private investment is worth more than one dollar socially. Therefore, the social values of a policy's benefits and costs depend on their effects on private investment. The net-social-benefit formula provides a correct basis for aggregating benefits and costs over time when dynamic efficiency is not attained, but the information required to calculate the net social benefit of an investment may not be available in practice.

If the effect on private investment of one dollar of benefits is equal to the effect of one dollar of costs, then the net social benefit of a policy is equal to its net present value, calculated using the social rate of time preference as the discount rate. Therefore, if the net-social-benefit formula cannot be calculated directly, policies might be evaluated using the net-present-value formula. However, the sensitivity of the results to alternative assumptions about the effects of benefits and costs on private investment should be investigated.

Notes

1. Strictly speaking, he could not consume at point A because he could not survive until next year if he consumed none of the commodity this year. This problem can be avoided by interpreting the vertical axis as being drawn at the minimal level of consumption this year that would allow survival.

2. It should be noted that social-indifference curves will reflect both the total quantities of goods and their distribution among individuals (Samuelson 1956).

3. If dynamic efficiency is attained, any part of $B_t - C_t$ that goes to investment can be assumed to earn the rate of interest, i. Therefore, the present value of the future changes in consumption caused by one dollar of investment will be equal to just one dollar, and the social value of one dollar of private investment will be equal to that of one dollar of consumption.

4. In practice, the real rate of interest is often approximated by subtracting the rate of inflation from the nominal interest rate, $i' = g - f$. From equation 4.23, $i = (g - f)/(1 + f)$. Therefore, $i' = i(1 + f)$, and the error of the approximation is small when f is small.

5. If the decision-making unit has a budget constraint that prevents it from undertaking all worthwhile projects, then the decision rule is to choose the combination of projects that has the largest net present value.

6. This problem could be avoided by calculating the benefit-cost ratio for the difference between the two investments and accepting the larger investment if this benefit-cost ratio is greater than one.

7. The ratios will be equal only if $D = B - C$, in which case they will both equal 1. If $D < B - C$, $(B - D)/C$ will be greater than $B/(C + D)$, and if $D > B - C$, $(B - D)/C$ will be less than $B/(C + D)$. Note that the net present value of the policy is not affected by the classification of the water pollution as a negative benefit or a positive cost, because $(B - D) - C = B - (C + D)$.

8. The number of positive real roots cannot exceed the number of changes in sign in net benefits. Therefore, an investment with a conventional net-benefit stream can not have more than one internal rate of return.

9. This problem might be avoided by calculating the internal rate of return for the difference between the investments being compared and applying rule III.A. However, because the distribution of net benefits corresponding to the difference between investments may be nonconventional even if the distributions for the investments themselves are conventional, this modified procedure might also fail.

10. For a synthesis of the debate sparked by Marglin's paper, see Sen (1967).

11. Interest-rate regulations are being phased out in the United States.

12. The discussion in this section is based on Bradford (1975) as corrected by Mendelsohn (1981). Boadway (1978) provides an analysis similar to, but more-general (and more-technical) than, that of Bradford. There is a vast literature on the subject of this section. Most of the earlier studies concentrated on the question of the appropriate discount rate to use when dynamic efficiency is not attained. The answers included the social rate of time preference (Arrow 1966; Kay 1972), the social opportunity-cost rate (Hirshleifer, Milliman, and Dehaven 1960; Mishan 1967) and a weighted average of the two (Harberger 1969; Sandmo and Dreze 1971). Marglin (1963b) and Feldstein (1964) are early examples of the more-general decision-rule approach taken in this section.

13. For simplicity, we assume that s, ρ, and γ, as well as the parameters θ and ϕ introduced later, are constant over time.

14. The assumptions of this special case underlie the conclusion by Arrow (1966) and Kay (1972) that public investments should be evaluated using net present value with the social rate of time preference as the discount rate.

5 Evaluation of Uncertain Benefits and Costs

The benefits and costs of air-pollution-control policies cannot be forecast with certainty because they will depend on the *state of the world* at each point in time in the future. For example, the benefits from installing a stack gas scrubber at a power plant may vary with atmospheric conditions, fuel use, operating efficiency, and so forth. Each combination of the possible values for these factors constitutes an alternative state of the world. Because future states of the world cannot be known in advance, benefit-cost analysis requires a method for evaluating uncertain benefits and costs.

This chapter discusses the economic theory of decision making under uncertainty and its implications for the evaluation of uncertain benefits and costs.[1] Recent challenges to the economic theory of decision making under uncertainty also are discussed.

Expected Monetary Value

The issues involved in evaluating policies under uncertainty can be illustrated by the simple example of an individual choosing among three mutually exclusive policies when there are only two possible states of the world. Table 5-1 shows the net benefits of each policy in each of the two possible states. Policy F clearly dominates policy G, because the net benefits of policy F are greater than or equal to those of policy G in all possible states. However, the appropriate ranking of policies F and H cannot be determined solely from the data in table 5-1, because policy F has larger net benefits in state 1 but smaller net benefits in state 2.

In order to be able to decide between policies F and H, the individual requires information on how likely each state is to occur. Given this infor-

**Table 5-1
Aggregation over States of the World**

	Net Benefits of Policy:		
State of the World	F	G	H
1	100,000	70,000	50,000
2	0	0	50,000

mation, one can express each policy in terms of a probability distribution of net benefits. For example, suppose that the probability of state 1 occurring is 0.6 and the probability of state 2 occurring is 0.4.[2] Policy F then corresponds to a probability of 0.6 of receiving $100,000 and a probability of 0.4 of receiving nothing, and policy H corresponds to a probability of 1.0 of receiving $50,000.

The individual's problem is to choose which probability distribution he prefers. One plausible set of decision rules for evaluating policies expressed as probability distributions is based on the concept of *expected monetary value* (EMV). The expected monetary value of a policy with n possible outcomes is defined as

$$\text{EMV} = \sum_{j=1}^{n} \Pi_j M_j \qquad (5.1)$$

where M_j is the dollar value of outcome j and Π_j is the probability that outcome j will occur. Thus the expected monetary value of a policy is equal to a weighted average of the policy's possible outcomes, with the weights being equal to the probabilities of the outcomes.

The decision rules based on expected monetary value are

Rule 1a. *A policy is worthwhile if its expected monetary value is greater than or equal to zero,*

and

Rule 1b. *In choosing among mutually exclusive policies, choose the one with the largest expected monetary value.*

Applying equation 5.1 to the data for policies F and H gives an expected monetary value of (0.6) $100,000 + (0.4) $0 = $60,000 for policy F and of (1.0) $50,000 = $50,000 for policy H. Therefore, by rule 1a, both policies are worthwhile; and, by rule 1b, policy F is preferable to policy H.

In effect, the expected-monetary-value decision rules recommend ignoring differences in the uncertainty of net benefits and evaluating policies as if their net benefits were certain to be equal to their expected monetary value (or mathematical expectation). The principal justification for using these rules is based on the "law of large numbers," which states that if a very large number of drawings were made from a probability distribution, the average value of the result would almost certainly be equal to the mathematical expectation. Therefore, if policy F were repeated many times, the average value of net benefits would almost certainly be $60,000. Since this is greater than the average net benefit that would result from repeating policy H, policy F should be chosen.

Evaluation of Uncertain Benefits and Costs

Since the law of large numbers essentially eliminates the uncertainty of average net benefits, it provides a strong rationale for using rules 1a and 1b to evaluate policies, *if policies with identical probability distributions are to be replicated many times.* However, most air-pollution-control policies do not satisfy this condition. Even when the same policy is applied in more than one case, the probability distribution of net benefits is likely to be different in each case as a result of differences between localities in atmospheric conditions, population density, and so forth. Therefore, it is necessary to consider the validity of the results obtained using the expected-monetary-value decision rules to evaluate policies for which the number of replications is small.

The classic demonstration that decision rules based on expected monetary values may lead to apparently incorrect decisions is the St. Petersburg Paradox. The paradox can be illustrated as follows. Suppose you are offered the opportunity to play the following game. An honest coin is to be tossed until it falls heads. If heads occurs for the first time on the kth toss, you receive 2^k dollars. The probability that the coin falls heads for the first time on the kth toss is $(1/2)^k$. Therefore, applying equation 5.1, the expected monetary value of the game is

$$\sum_{k=1}^{\infty} 2^k \left(\frac{1}{2}\right)^k$$

which is infinite. This implies that a person basing his decisions on expected monetary values would choose to play this game rather than accept a certain payment of any finite amount of money. Introspection suggests that there is some amount of money, received for certain, that would be preferred to the game, and therefore that decisions based on expected monetary value may be incorrect.

The data for policies F and H in table 5-1 provide a more relevant, if less dramatic, demonstration that a rational individual would not necessarily choose the probability distribution with the largest expected monetary value. Policy F has a larger expected monetary value than policy H but also involves more uncertainty. If the chosen policy could be carried out only once, a rational individual might choose policy H because he felt that the larger expected monetary value of policy F was not sufficient to compensate him for its greater uncertainty.

Expected Utility

These examples suggest that the expected-monetary-value decision rules do not always coincide with the actual decision rules used by rational individuals. An alternative set of decision rules that is capable of explaining these examples is based on the concept of *expected utility*. Expected utility is defined as

$$\sum_{j=0}^{n} \Pi_j U_j(M_j) \qquad (5.2)$$

where $U_j(M_j)$ is the utility the individual receives from the net benefits of outcome j. The decision rules based on expected utility are

Rule 2a. *A policy is worthwhile if its expected utility is greater than or equal to zero.*

and

Rule 2b. *In choosing among mutually exclusive policies, choose the one with the largest expected utility.*

At first glance, the expected-utility decision rules seem to share with the expected-monetary-value rules the drawback of not reflecting an individual's attitudes toward uncertainty. However, von Neumann and Morgenstern (1947) have shown that, given certain "postulates of rational choice," there is a way of assigning utility values to outcomes such that the utility of any uncertain situation equals the expected value of the utilities of the possible outcomes.[3] Therefore, an individual whose preferences satisfy the rationality postulates will be indifferent between two uncertain situations with the same expected utility, even if the distributions of possible outcomes are very different.

The method used to assign utility values to outcomes will be illustrated by constructing a utility function for a hypothetical individual. The first step is to choose two amounts of income, say, $0 and $100,000, and assign them arbitrary utility values, say, 0 and 100, respectively.[4] The utility, for decisions under uncertainty, of a third amount of income is then determined by asking the individual to choose the amount of money such that he would be indifferent between receiving that amount for certain or owning a lottery offering a 0.5 probability of receiving $0 and a 0.5 probability of receiving $100,000.

Suppose that the individual says that he would be indifferent between receiving $30,000 for certain and the lottery; that is, $30,000 is the *certainty equivalent* of the lottery. Since he is indifferent between $30,000 and the lottery, their utilities must be the same;

$$U(\$30,000) = (0.5) \, U(\$0) + (0.5) \, U(\$100,000)$$
$$= (0.5) \, 0 + (0.5) \, 100 = 50$$

Evaluation of Uncertain Benefits and Costs

Therefore, the utility of $30,000 is assigned a value of 50. The utility of other amounts of income can be determined by repeating the procedure. For example, if the individual would be indifferent between $56,000 and a lottery with probability 0.5 of $30,000 and 0.5 of $100,000, then:

$$U(\$56,000) = (0.5)\, U(\$30,000) + (0.5)\, U(\$100,000)$$
$$= (0.5)\, 50 + (0.5)\, 100 = 75.$$

The individual's complete utility function for decisions under uncertainty is shown in figure 5-1. The concave shape of the utility function shows that total utility increases less fast than total income, which implies that the marginal utility of income decreases as income increases.

Given the individual's utility function, equation 5.2 can be calculated for any probability distribution of outcomes; and decision rules 2a and 2b can be used to evaluate policies without further consultation with the individual. For example, the expected utility of policy F is equal to (0.6) $U(\$100,000)$ + (0.4) $U(\$0)$ = 60; the expected utility of policy H is $U(\$50,000)$ = 71. Therefore, the individual would prefer policy H, despite the higher expected monetary value of policy F.

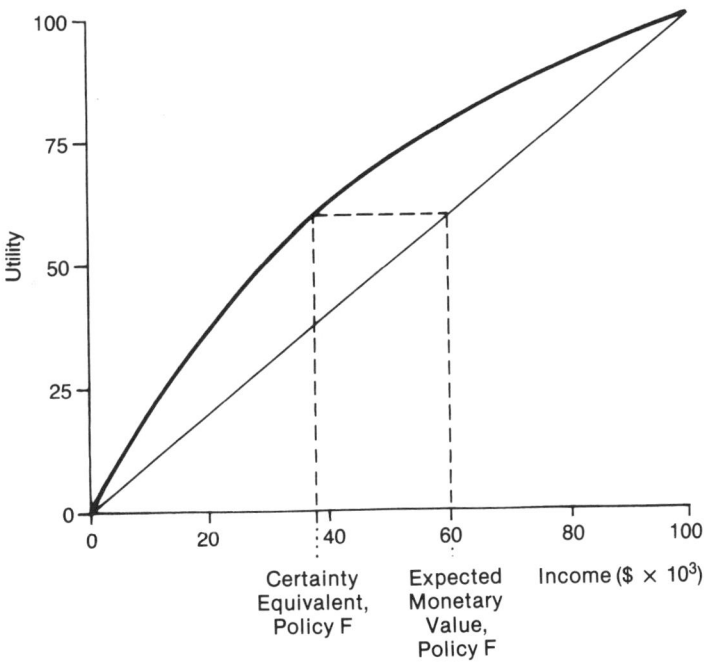

Figure 5-1. Utility Function for Decisions Involving Uncertainty

For an individual with a concave utility function, the certainty equivalent of any uncertain situation will be less than its expected monetary value. The certainty equivalent is found by calculating the expected utility of the probability distribution of income in the uncertain situation and determining the amount of income, received for certain, that has the same utility. For example, assuming for the moment that the individual's total income is equal to the net benefits of the policy, the certainty equivalent of policy F is $38,000, because the expected utility of policy F is 60 and, from figure 5-1, 60 = $U(\$38,000)$.

An individual with a concave utility function is said to be *risk averse*. He would always prefer a certain income to a probability distribution having the same expected monetary value. Therefore, he would not gamble at fair odds,[5] and would pay more than the expected loss to insure against risk.[6] He would not gamble at fair odds because the value to him of the gamble (that is, its certainty equivalent) would be less than its expected monetary value. To illustrate that he would pay more than the expected loss to insure against risk, suppose that the individual owns an asset worth $100,000 and that there is a probability of 0.4 that the asset will be lost. If he did not insure, he would have a probability 0.6 of $100,000 and probability 0.4 of $0. If he bought insurance at a cost of D dollars, he would have $(100,000 - D)$ for certain. As shown in figure 5-1, the certainty equivalent of the probability distribution if he did not insure would be $38,000. Therefore, the individual would be willing to pay up to $62,000 for insurance even though the expected monetary value of the loss is only $40,000.

If the utility function for an individual were linear, the certainty equivalent of a probability distribution would be equal to its expected monetary value. The individual would be *risk neutral*, being just willing to gamble at fair odds and being unwilling to pay more than the expected loss to insure against risk. If an individual had a convex utility function, the certainty equivalent of a probability distribution would be greater than its expected monetary value. The individual would prefer risk, accepting gambles with unfavorable odds and being unwilling to buy insurance unless the premium were less than the expected loss.

The shape of the utility function clearly can have a critical effect on the choice of policies using decision rules 2a and 2b. Utility functions for decisions under uncertainty are usually assumed to be concave. The principal empirical evidence in favor of this hypothesis is that most individuals prefer to hold a variety of assets. If individuals were risk neutral or preferred risk, they would instead commit their total wealth in the single asset that offered the highest mathematical expectation of return, no matter how risky it was. However, concave utility functions do not explain the observed willingness of many individuals to accept small gambles at unfavorable odds.

Friedman and Savage (1948) noted that some individuals are willing

Evaluation of Uncertain Benefits and Costs

both to buy insurance and to gamble at unfavorable odds, suggesting that their utility functions are concave at low levels of income and convex at high levels of income. However, such individuals would not remain in a convex range of their utility functions for long, because their preference for risk would lead them to accept all gambles whose odds were not excessively unfavorable. Therefore, economists generally assume that utility functions are concave and explain the gambling at unfavorable odds that does occur as primarily recreational in nature (Hirshleifer and Riley 1979).

Alternatively, gambling at unfavorable odds and other apparent anomalies in actual decision making under uncertainty may indicate that individual preferences do not satisfy the postulates of rational choice. This possibility, originally argued by Allais (1953), has received considerable attention from experimental psychologists in recent years.

Prospect Theory

Kahneman and Tversky (1979) have produced evidence that decisions made under uncertainty are not consistent with the maximization of expected utility. The evidence is based on the responses of individuals in hypothetical choice situations. For example, they asked 95 individuals to choose between alternative A, a lottery with probability 0.80 of winning 4,000 and probability 0.20 of winning nothing, and alternative B, 3,000 to be received for certain.[7] Eighty percent of the respondents chose alternative B. In a second problem, the same individuals were asked to choose between alternative C, a lottery with probability 0.20 of winning 4,000 and 0.80 of winning nothing, and alternative D, a lottery with probability of 0.25 of winning 3,000 and of 0.75 of winning nothing. Sixty-five percent chose C.

The choice of alternative B in the first problem and of C in the second problem is inconsistent with expected utility theory. To see this, let $U(0) = 0$. Then $U(A) = (0.80) \, U(4,000)$ and $U(B) = U(3,000)$. Since B is chosen over A,

$$(0.80) \, U(4,000) < U(3,000) \qquad (5.3)$$

Similarly, the choice of C over D implies

$$(0.20) \, U(4,000) > (0.25) \, U(3,000) \qquad (5.4)$$

However, multiplying both sides of inequality 5.4 by 4.0 yields

$$(0.80) \, U(4,000) > U(3,000)$$

which is inconsistent with inequality 5.3.

Kahneman and Tversky present a number of such observations and conclude that expected-utility theory is not an adequate descriptive model of decision making under uncertainty. They propose in its place a descriptive model, called *prospect theory,* which modifies expected utility theory in order to accommodate these observations. First, outcomes are expressed as gains or losses from a neutral reference point, and the value function for outcomes is specified to be concave for gains and convex for losses. Second, the value of an uncertain outcome is multiplied by a decision weight, which is a monotonic function of the probability of the outcome, rather than by the probability itself. The decision weights are assumed to be larger than the corresponding probabilities for low probabilities and smaller than the probabilities for large probabilities.

Prospect theory appears to be a promising descriptive model of decision making under uncertainty, although it has not yet been tested extensively using actual as opposed to hypothetical choices.[8] Whether or not prospect theory turns out to be a better *descriptive* model for individuals, expected-utility theory is likely to remain a better *normative* model for social decision making under uncertainty, because an individual whose preferences satisfied the postulates of prospect theory would make irrational choices. For example, the choice between two alternatives could be reversed simply by changing the wording used to describe the alternatives. This problem does not arise if choices are based on expected-utility theory. Since coherency of preferences is an important criterion for public-policy decisions, expected-utility theory appears to provide a preferable basis for social decision making under uncertainty.

Implications for Policy Evaluation

We have considered three theoretical bases for decision making under uncertainty: expected monetary value, expected utility, and prospect theory. We concluded that expected-utility theory is the most appropriate basis for social decision making under uncertainty and, therefore, that decision rules 2a and 2b are the appropriate rules to use in aggregating over states of the world. However, application of these decision rules requires knowledge of the utility functions of the individuals affected by a policy, knowledge that generally will not be available. Therefore, it is useful to consider the magnitude of the errors that are likely to occur if the expected-monetary-value decision rules, 1a and 1b, are used to evaluate policies under uncertainty.

Decision rules 1a and 1b would be equivalent to decision rules 2a and 2b if utility functions for decisions under uncertainty were linear. To see this,

Evaluation of Uncertain Benefits and Costs

let $U(0) = 0$. Then $U_j(M_j)$ would equal kM_j where k is the (constant) marginal utility of income. Substituting in 5.2,

$$\sum_j \Pi_j U_j(M_j) = k\sum_j \Pi_j M_j$$

Since expected utility would be proportional to expected monetary value, rules 1a and 1b would result in the same choices as rules 2a and 2b. The equivalence of the expected-monetary-value and expected-utility decision rules in this case also can be seen by noting that the certainty equivalent of a probability distribution is equal to its expected monetary value when the utility function for decisions under uncertainty is linear.

Therefore, if the individuals affected by a policy were risk neutral, no errors would be introduced by using rules 1a and 1b in place of 2a and 2b. However, individuals are not in general risk neutral; therefore, the expected-monetary-value decision rules will not in general be equivalent to the expected-utility rules.

The magnitude of the errors introduced by using rules 1a and 1b will be positively related to the degree of nonlinearity of the utility function, the dollar values of the policies' outcomes, and the degree of interdependence between the outcomes and other sources of income. These relationships are illustrated in figure 5-2. To simplify the discussion, it will be assumed first that the outcomes of a policy are independent of other sources of income. This condition will be satisfied if the level of income from sources other than the policy is not affected by the states of the world relevant for the policy. For example, if the net benefits of an air-pollution-control policy depended only on atmospheric conditions, there would be no correlation between its net benefits and other sources of income for most individuals.

Suppose that income from other sources is Y_1, and that the net benefit of the policy in some state of the world is equal to M. The marginal utility of income when income is Y_1, which we write $U'(Y_1)$, is equal to the slope of the utility function at Y_1, which in turn is equal to the slope of line LL tangent to the utility function at Y_1. Treating the marginal utility of income as a constant is equivalent to using line LL as an approximation to the utility function in this range. The increase in utility from receiving M would be calculated as $U'(Y_1)M$. The actual increase in utility if M is received is equal to

$$U_1(Y_1 + M) - U_1(Y_1).$$

As shown in figure 5-2,

$$U'(Y_1)M > U_1(Y_1 + M) - U_1(Y_1).$$

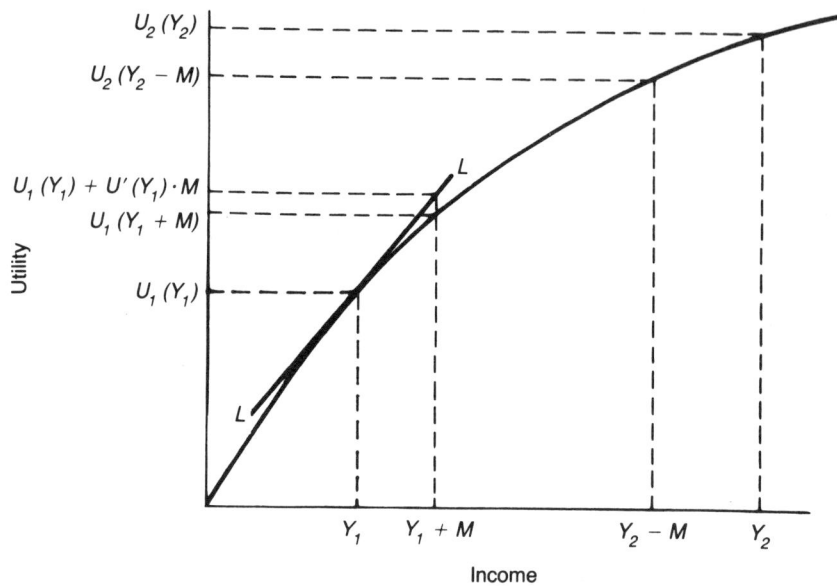

Figure 5-2. Approximation Errors Using Expected-Monetary-Value Decision Rules.

Therefore, the assumption of constant marginal utility results in an overestimate of the value of M. The magnitude of the error will clearly be smaller the more linear the utility function and the smaller the size of M relative to Y_1. Thus, if the outcomes of the policies being compared are small and independent of other sources of income, the use of decision rules 1a and 1b will not result in large errors unless individuals are extremely risk averse.

However, this result does not necessarily hold if there is interdependence between the outcomes of a policy and other sources of income. To illustrate, suppose that there are two relevant states of the world and that income from other sources is Y_1 in state 1 and Y_2 in state 2, with $Y_2 > Y_1$. The concave shape of the utility function implies that the marginal utility of income will be lower in state 2 than in state 1. Therefore, the expected utility of a policy will not be proportional to its expected monetary value, even if the net benefits of the policies are small relative to income.

For example, if a policy has net benefits of M in state 1 and of $-M$ in state 2, and the states are equally likely to occur, $\Pi_1 = \Pi_2$, the expected monetary value will be 0, $\Pi_1 M - \Pi_2 M = 0$. However, expected utility will be greater than 0. As shown in figure 5-2, the value of M in state 1, $U_1(Y_1 + M) - U_1(Y_1)$, is greater in absolute magnitude than the value of $-M$ in state 2, $U_2(Y_2) - U_2(Y_2 - M)$. Therefore,[9]

$$\Pi_1 U_1(M) - \Pi_2 U_2(M) > 0. \tag{5.5}$$

In summary, the use of decision rules based on expected monetary value will not result in large errors relative to the use of expected-utility decision rules if the outcomes of the policies being compared are independent of, and small relative to, other sources of income. If the conditions on the size and independence of policy outcomes are not satisfied, then the use of the expected-monetary-value decision rules may result in incorrect policy choices. It is important to note that the expected monetary value may understate, rather than overstate, the value of an uncertain situation. If individuals are risk averse, then the expected monetary value will understate the amount they would be willing to pay for a policy with uncertain outcomes, if the outcomes of the policy are negatively correlated with income from other sources, *or* if the policy reduces the overall level of uncertainty.

The finding that expected monetary value may understate the value of a policy that reduces the overall level of uncertainty is particularly relevant for air-pollution-control policies. A principal reason that the benefits of air-pollution-control policies are uncertain is that they consist of the reduction of uncertain damages. Therefore, air-pollution-control policies are similar in effect to the purchase of insurance against potential losses. As noted previously, risk-averse individuals will be willing to pay more than the expected monetary value of losses to insure against risk. Since the expected monetary value of the benefits of an air-pollution-control policy is equal to the expected monetary value of the losses averted, the expected monetary value of the benefits may understate the amount that individuals would be willing to pay for the benefits of air-pollution control.

The possibility that expected monetary value may understate rather than overstate the value to risk-averse individuals of a policy with uncertain outcomes is generally overlooked in discussions of policy evaluation under uncertainty. As a result, it is often recommended that the expected monetary value of the outcomes be used as the measure of net benefits in each year and that a "risk premium" be included in the discount rate to allow for risk aversion. The net present value would then be calculated as

$$\sum_{t=0}^{T} \frac{1}{(1 + \gamma + p)} \text{EMV}_t \tag{5.6}$$

where p is the risk premium and EMV_t is the expected monetary value of net benefits in year t.

If expected monetary value were an understatement of the value of net benefits for the policy being evaluated, then the use of a risk premium would have the unfortunate effect of increasing the size of the error. Even if expected monetary value were known to overstate the value of net benefits, the use of a risk premium would not be a very desirable procedure because the size of the adjustment for risk would increase over time as a result of compounding, whether or not the degree of uncertainty increased.

A preferable approach is to evaluate the relationship between expected monetary value and the value of net benefits directly. In this process, benefits and costs should be considered separately, because the direction and magnitude of the effects of risk aversion may be different for each.

Expected monetary value will generally understate the value of uncertain costs to risk-averse individuals, but the magnitude of the total error may be small. This is because the final incidence of costs is likely to be spread over many individuals (for example, through changes in government tax revenues). Therefore, the effect on each individual is likely to be small relative to his income. As noted previously, when the effects are small and are independent of other sources of income, the use of expected monetary value generally will not result in a large error for an individual. It can be shown that the error for each individual decreases faster than the number of individuals increases, so that the total error (that is, the product of the error for each individual times the number of individuals) decreases as the number of individuals increases, and becomes zero in the limit (Arrow and Lind 1970).[10]

As discussed previously, expected monetary value may either overstate or understate the value of uncertain benefits to risk-averse individuals. Risk spreading may again reduce the size of the error. However, since each individual's benefits from clean air do not necessarily decrease as the number of individuals affected increases, risk spreading may reduce the total error from using expected monetary value less for benefits than for costs (Fisher 1973). If the total error is larger for benefits than for costs, then net benefits (that is, benefits minus costs) will be understated by expected monetary value if benefits are understated.

Summary

Three theoretical bases for decision making under uncertainty have been discussed: expected monetary value, expected utility, and prospect theory. Expected-utility theory provides an appropriate basis for evaluating uncertain benefits and costs, but the information required to calculate expected utility will generally not be available in practice. Calculation of expected monetary values is more feasible, but the expected value of net benefits will generally not be equal to the true value of the policy to those affected by it.

Because expected monetary value may either overstate or understate the value of net benefits, inclusion of a risk premium in the discount rate is not an appropriate procedure to use in adjusting for uncertainty. Since the

Evaluation of Uncertain Benefits and Costs 75

actual errors may be small when many individuals are affected, the recommended procedure for evaluating uncertain benefits and costs is to use the expected monetary value of net benefits in each year, unless there is strong evidence that it either overstates or understates the value of net benefits. If such evidence exists, the expected monetary value should be adjusted directly, rather than through the use of a (positive or negative) risk premium. If the direction of error is known, but there is insufficient information to determine the appropriate size of the adjustment, then the expected monetary value can be interpreted as an upper or lower bound for the value of net benefits.

Notes

1. Hirshleifer and Riley (1979) provide an excellent nontechnical survey of the economics of uncertainty.

2. The probability of each state is represented by a number not less than 0 or greater than 1 and the sum of the probabilities of all states is equal to 1. Evaluation of the probability of each state may be based on the relative frequency with which it has occurred in the past, expert opinion, simulation, and so forth.

3. The postulates of rational choice underlying the use of expected-utility decision rules are discussed in Friedman and Savage (1948) and Luce and Raiffa (1957).

4. Arbitrary utility values can be used for these two amounts because the utility function need be unique only up to a positive linear transformation.

5. The odds are said to be fair if the expected value of the outcomes is equal to the amount bet, so that the mathematical expectation of net returns is 0.

6. As discussed later on, these conclusions may not hold if the outcomes are not independent of other sources on income.

7. The outcomes refer to Israeli pounds. The median net monthly family income was about 3,000 Israeli pounds.

8. Some evidence based on actual choices is presented in Tversky and Kahneman (1981).

9. Inequality 5.5 implies that a risk-averse individual may nevertheless accept a gamble at fair odds, or even at somewhat unfavorable odds, if the outcomes of the gamble are negatively correlated with his other sources of income.

10. When many air-pollution policies are undertaken, the resulting pooling of risks may also decrease the effects of any errors arising from the use of expected monetary value (James 1975).

6

Valuation of Priced and Unpriced Commodities

Air-pollution-control policies affect social well-being by changing the quantities of commodities consumed and produced. The number of commodities affected by a policy (including environmental quality as well as other goods and services) is typically very large. Therefore, benefit-cost analysis of air-pollution control requires a method for aggregating over changes in many diverse types of commodities. Because the goal is to obtain aggregate measures of the net benefits that individuals obtain from the changes, the values assigned to commodities should reflect their value to the individuals affected. Commodity markets provide a valuable source of information on these values.

This chapter discusses the use of market data to estimate the dollar values of the benefits and costs of a policy. Generally, the market data available for estimating costs will be more complete than the data for benefits because prices generally will exist for the commodities representing the costs but often will not exist for the commodities representing the benefits. The chapter discusses the valuation of changes in priced commodities when markets are perfectly competitive and when they are imperfectly competitive. The use of market data for related commodities to value changes in commodities for which markets do not exist is also discussed.

Perfectly Competitive Markets

In perfectly competitive markets, consumers and producers face the same prices for commodities and both act as if they cannot affect prices by their own actions. A consumer will purchase a commodity up to the point where the dollar value to him of the last unit consumed is equal to its price, and a producer will supply a commodity up to the point where the cost to him of the last unit sold is equal to its price. As a result, the market price of a commodity will be equal to its marginal dollar value in consumption and to its marginal cost of production.

Therefore, when aggregating over commodities that are traded in perfectly competitive markets, the appropriate value to place on each commodity is its market price. However, if the policy being evaluated would result in a large change in the quantity of a commodity consumed or produced, the market price of the commodity might be affected. Therefore,

Consumer's Surplus

The valuation of a large change in consumption is illustrated in figure 6-1 for a single individual. The individual's demand curve shows the quantity of the commodity he would purchase at each price. For example, he would purchase Q_1 units of the commodity if the price per unit were P_1, Q_2 units if the price were P_2, and 0 units if the price were P_3 or higher. Because the individual will purchase a commodity up to the point where the dollar value to him of the last unit purchased is equal to its price, the demand curve also indicates the value to him of each unit. Thus the dollar value of unit Q_1 is P_1, the value of unit Q_2 is P_2, and the value of the first unit is slightly less than P_3.

Since the height of the individual's demand curve is equal to the value to him of each unit, the total dollar value to him of a given number of units is equal to the area under his demand curve for that number of units. For

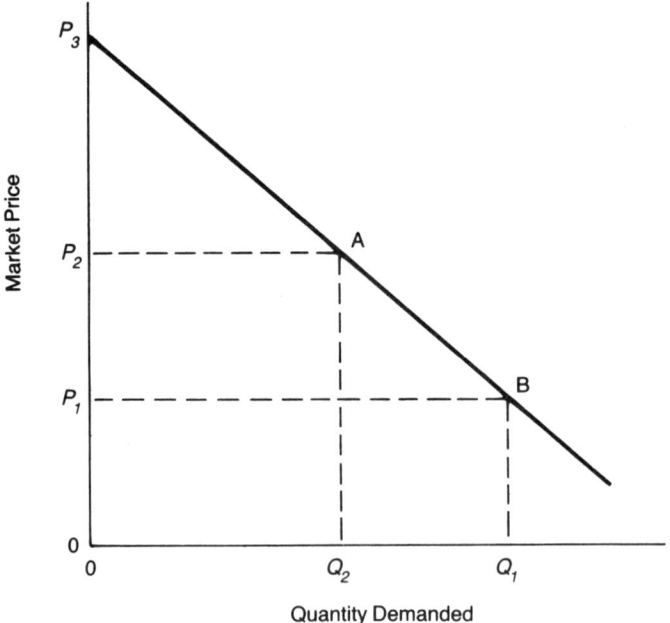

Figure 6-1. Individual's Demand Curve for a Commodity

example, the total value of Q_2 units is equal to the area $0P_3AQ_2$. However, if the commodity could be purchased at price P_2, the total cost of Q_2 units to the consumer would be only the area $0P_2AQ_2$. Since the total cost would be less than the total value to the consumer, he would realize a net gain equal to the difference, area P_2P_3A. The difference between the total value and the total cost of a commodity to a consumer is referred to as his *consumer's surplus*.

If the price of a commodity changes, the consumer's surplus also will change; and the change in consumer's surplus can be interpreted as the dollar value to the consumer of the change in price. For example, if the price decreased from P_2 to P_1, then the individual would increase his purchases from Q_2 to Q_1. The consumer's surplus from purchasing Q_1 units at price P_1 would be equal to area P_1P_3B. Therefore, a decrease in the price from P_2 to P_1 would increase his consumer's surplus by an amount equal to area P_1P_2AB; and this area provides an appropriate measure of the amount by which he would benefit if the price fell from P_2 to P_1.

It should be noted that there is an extensive theoretical literature questioning the use of this measure of a consumer's benefit from a price change. Alternative measures that have been proposed include the "equivalent variation" and the "compensating variation." The controversy appears finally to have been settled for practical purposes by Willig (1976), who showed that consumer's surplus provides an adequate approximation to both the equivalent and compensating variations for most cases likely to arise in practice.[1]

Producer's Surplus

The value of a large change in the quantity of a commodity produced will be equal to the sum of the values of the resulting changes in the quantities of the inputs (for example, labor and capital) required to produce the commodity. The valuation of a large change in an input is illustrated in figure 6-2 for a single individual's supply of labor. The individual's supply curve for labor shows the quantity of labor he would supply at each wage rate. For example, he would supply H_3 hours of labor if the wage rate were W_3, H_2 hours if the wage rate were W_2, and 0 hours if the wage rate were W_1 or lower.

An individual will be willing to supply labor up to the point where the cost to him of the last hour supplied is equal to the wage rate. The cost of supplying each hour of labor is the dollar value to him of the activities he has to give up in order to spend that hour working.[2] Therefore, the supply curve indicates the opportunity cost of each hour of labor supplied. Thus the opportunity cost of hour H_3 is W_3, the opportunity cost of hour H_2 is

W_2, and the opportunity cost of the first hour worked is slightly greater than W_1.

Since the height of the individual's supply curve is equal to the opportunity cost of each hour worked, the total opportunity cost of a given number of hours is the area under his supply curve for that number of hours. For example, his total opportunity cost of H_2 hours of labor is equal to area $0W_1AH_2$.

If the wage rate were W_2, the total wages he would receive would be $0W_2AH_2$. Since the total opportunity cost to him of the labor supplied is less than the total wages received, he would realize a net gain equal to the difference, area W_1W_2A. This measure of his net gain is referred to as his *producer's surplus*. If the wage rate increased to W_3, his producer's surplus would increase by area W_2W_3BA; and this provides an appropriate measure of the dollar value of the benefit he would receive from an increase in the wage rate from W_2 to W_3.[3]

Aggregate Consumers' and Producers' Surplus

Consumer's and producer's surpluses for individuals generally cannot be calculated directly because sufficient information is not available on individual demand curves for commodities and supply curves for inputs. At best, information will be available on aggregate demand curves and supply curves for markets, and the information on supply curves generally will refer to commodities rather than to inputs. Therefore, it is necessary to consider the relationship between the aforementioned measures and measures based on market demand and supply curves for commodities.

The market demand curve for a commodity is simply the horizontal summation (that is, along the quantity axis) of the demand curves of each individual. Therefore, the height of the market demand curve remains equal to the dollar value of each unit. The aggregate dollar value of a given quantity of a commodity will be equal to the area under the market demand curve for that quantity, and the aggregate consumers' surplus resulting from a price change can be measured in a way analogous to the measurement in the case of a single consumer.

Similarly, the market supply curve for an input is the horizontal summation of the supply curves for each individual. Therefore, the height of the market supply curve for an input is equal to the opportunity cost of each unit of the input. Inputs are combined into commodities by firms. Under conditions of perfect competition, a firm's supply curve for a commodity corresponds to the portion of its marginal-cost curve lying above its average variable-cost curve.[4] The marginal cost of each unit will be equal to the dollar value of the inputs required to produce that unit, which in turn is

Priced and Unpriced Commodities

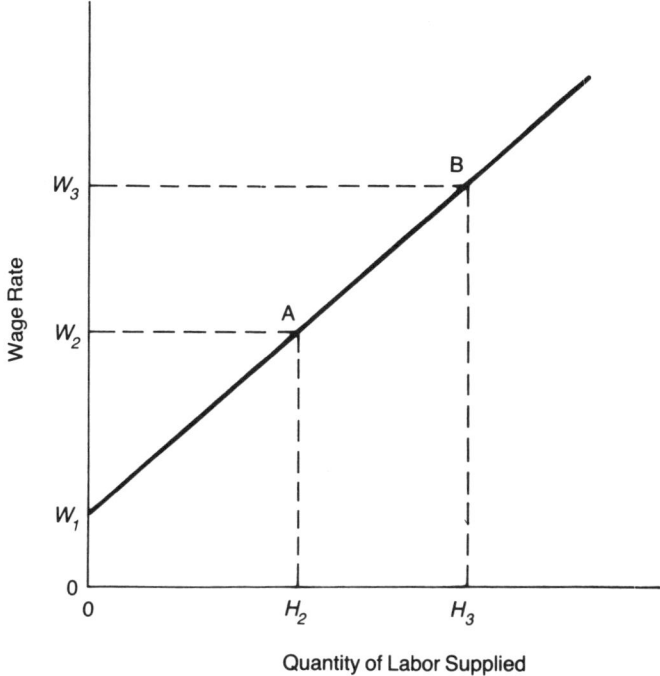

Figure 6-2. Individual's Labor-Supply Curve

equal to the opportunity cost of the inputs. Therefore, the height of a firm's supply curve indicates the dollar value of the marginal opportunity cost of each unit of the commodity.

Market supply curves for commodities are obtained by aggregation of the supply curves of all the firms in the industry. The market supply curve for a commodity generally will not be equal to the horizontal summation of each firm's supply curve because the prices of inputs may increase when all firms in the industry expand output. This may occur as a result of the upward slope of the market supply curves for inputs, or because inputs have to be bid away from other industries. In either case, the price paid for an input by an industry will reflect the opportunity cost of the last unit purchased; and the area under the market supply curve for the commodity will be equal to the total opportunity cost of the inputs used by the industry. Therefore, the difference between the total revenue received by an industry and the area under the market supply curve for the commodity is equal to the total producers' supply received by suppliers of inputs to that industry.

Aggregate consumers' and producers' surpluses are illustrated in figure 6-3, which shows the market demand and supply curves for a commodity together with the market price, P, and total quantity transacted, Q. The

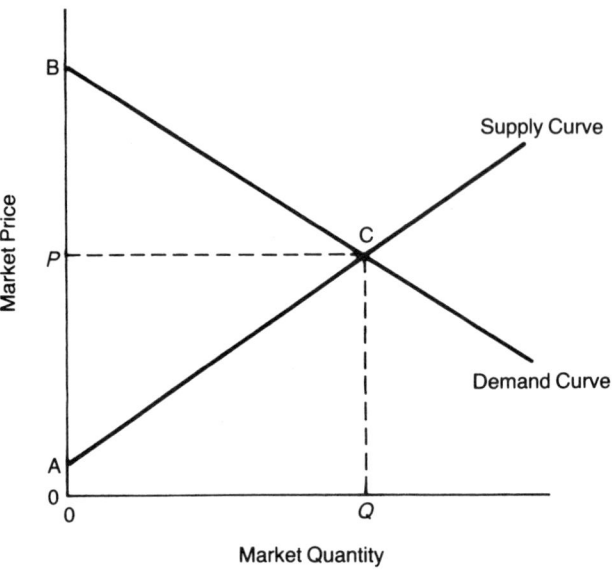

Figure 6-3. Market Demand and Supply Curves for a Commodity

aggregate consumers' surplus is equal to the area under the market demand curve and above the market price line, area PCB. Aggregate producers' surplus is equal to the area above the market supply curve and below the price line, area PCA.

To illustrate the valuation of a large change in the production and consumption of a commodity, suppose that the implementation of an air-pollution-control policy would require using N units of a commodity. As shown in figure 6-4(a), implementation of the policy would shift the demand curve for the commodity to the right by N units (distance AC) from curve D_0 to curve D_1. The price of the commodity would increase from P_0 to P_1, the quantity produced would increase from Q_0 to Q_S, and the quantity consumed privately would decrease from Q_0 to Q_D. The dollar value of the decrease in private consumption is equal to area $Q_D ABQ_0$, and the dollar value of the opportunity costs of the increase in production is $Q_0 BCQ_S$. The total dollar value is thus $Q_D ABCQ_S$.

If the total dollar value were estimated as the original price, P_0, times N units, the result would be an underestimate of the dollar value, whereas the use of the new price, P_1, would result in an overestimate. If detailed information on the demand and supply curves were not available, but the effect on the price of the commodity could be estimated, the total dollar value of N units of the commodity could be approximated by using an average of the original and new prices.

Priced and Unpriced Commodities

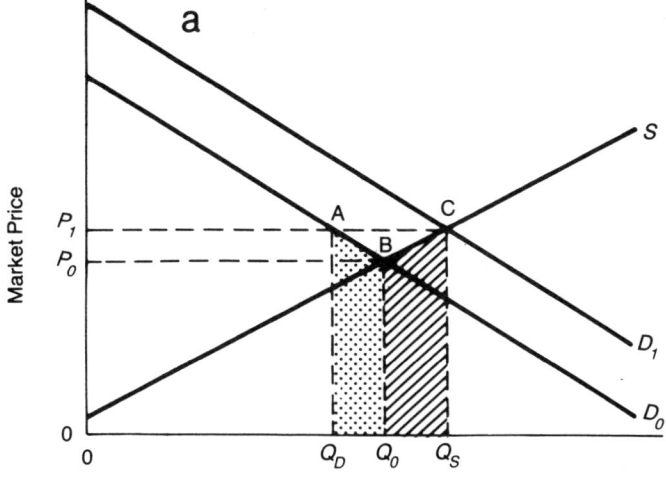

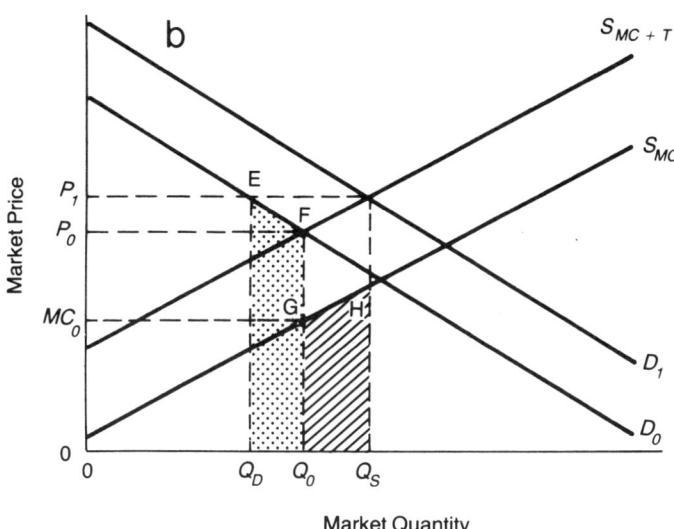

Figure 6-4. Valuation of Large Changes

Imperfectly Competitive Markets

The preceding section discussed procedures for evaluating the dollar value of changes in the quantities of commodities when markets are perfectly competitive. The adjustments to these procedures that are necessary when markets are not perfectly competitive are illustrated in this section by considering the effects of commodity taxes on the calculation of dollar values.

A commodity tax will result in a divergence between the price paid by consumers and the marginal cost of production. This is illustrated in figure 6-4(b) for the case of a constant tax of T dollars per unit levied on the suppliers of a commodity. The supply curve in the absence of the tax would be equal to the industry's marginal cost curve, which is labeled S_{MC} in the diagram. Given the tax, the cost to suppliers of a unit of the commodity will be increased by the amount of the tax (distance FG). The supply curve with the tax is labeled S_{MC+T}, and the market price, P_0, is determined by the intersection of this supply curve and the demand curve. The dollar value to consumers of the last unit of the commodity is equal to the price inclusive of tax, P_0, whereas the marginal cost is equal to the price exclusive of tax, $P_0 - T = MC_0$.

To illustrate the calculation of the dollar value of a large change in this commodity, again suppose that an air-pollution-control policy required the use of N units of the commodity. As shown in figure 6-4(b), implementation of the policy would result in a shift in the demand curve from curve D_0 to curve D_1. The market price inclusive of tax would increase from P_0 to P_1, private consumption would decrease from Q_0 to Q_D, and production would increase from Q_0 to Q_S. The total dollar value of the N units would be equal to the dollar value of the change in consumption, area $Q_D EF Q_0$, plus the dollar value of the change in production, area $Q_0 GH Q_S$.

As this example makes clear, the existence of market imperfections complicates the calculation of the dollar value of changes in the quantities of commodities resulting from air-pollution-control policies. In the case of a commodity tax, calculation of the dollar value of a large change requires knowledge of both the demand and the supply curves for the commodity. The use of an average of the original and new prices will not provide nearly as good an approximation of the dollar value as in the case of perfect competition.

Commodity taxes are just one of many possible sources of market imperfections. Markets will also be imperfect if firms are large enough to affect the prices they receive for their output or pay for their inputs, if unions are able to affect the wage rate, if consumers have incomplete information on market prices, and so on. Therefore, the conditions for perfect competition are unlikely to be satisfied for most commodities, and the prices paid by consumers generally will not be equal to the marginal cost of production. Therefore, market prices will not be fully accurate measures of the values of commodities, even for small changes. However, attempts to obtain better measures for benefit-cost analyses should take into account the cost of the extra information required. If the probable adjustments to market prices are unlikely to affect the final choice of policies, then the

Priced and Unpriced Commodities 85

investment of time and effort in acquiring the additional information required for the adjustments may not be worthwhile.[5]

Nonexistent Markets

A critical problem in performing benefit-cost analyses of air-pollution-control policies is that markets generally do not exist for environmental quality. As a result, the assignment of dollar values to changes in environmental quality requires either the use of market data for related commodities or the use of nonmarket data such as the results of surveys. Freeman (1979b) presents an excellent review and critique of the theoretical basis for a number of procedures that have been, or could be, used. Studies using some of these approaches will be discussed in the following chapters. In this chapter, the method that holds the most promise for deriving useful results from the type of market data usually available will be discussed.

Use of Data for Affected Commodities

Environmental quality can be viewed as an unpriced input to the production processes for commodities. Changes in environmental quality will affect the amount of output obtained from given amounts of other inputs, and therefore will shift the supply curves for the affected commodities. If market data exist for an affected commodity, the dollar value of the benefits resulting from a shift in the supply curve for that commodity can be calculated using the concepts of consumers' and producers' surplus discussed previously.

To illustrate, suppose that an air-pollution-control policy would result in a decrease in ambient concentrations of ozone in a tomato-growing area and that this decrease would result in an increase in the output of tomatoes using given quantities of other inputs. The result would be a decrease in marginal cost for each level of output and hence a shift to the right in the supply curve for tomatoes.

The effect on total consumers' and producers' surplus is illustrated in figure 6-5. The supply curve for tomatoes before the decrease in ozone is labeled S_0, and the total consumers' and producers' surplus given this supply curve is equal to area AFB_0. The supply curve after the decrease in ozone is labeled S_1. Total consumers' and producers' surplus given supply curve S_1 is equal to AGB_1. Thus a shift in the supply curve from S_0 to S_1 would result in an increase in total consumers' and producers' surplus from this commodity equal to the shaded area, B_0FGB_1. Similarly, if the supply curve shifted from S_1 to S_0 as a result of increased air pollution, the decrease

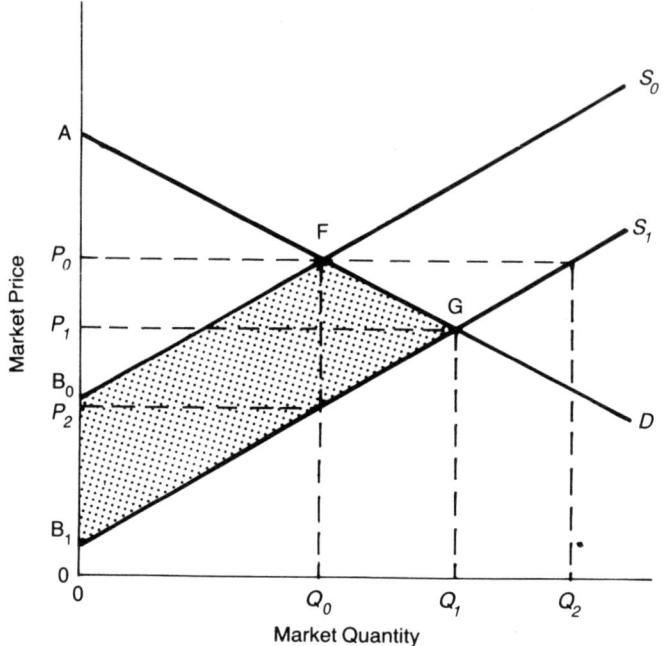

Figure 6-5. Effect of Improvement in Environmental Quality

in total consumers' and producers' surplus from this commodity also would be area B_0FGB_1.

If adequate information on demand and supply curves were available for all affected commodities, the total change in consumers' and producers' surplus from a change in air pollution could be calculated by summing the changes in consumers' and producers' surplus for each commodity. This would provide an appropriate measure of the value of the benefits to society of the change in air pollution if the distribution of benefits across individuals did not affect social well-being. However, if marginal social utilities of income are not equal, then the distribrution of changes in consumers' and producers' surplus must be taken into account.

In order to incorporate fully distributional considerations into the analysis, it would be necessary to measure gains and losses at the individual, rather than the market, level. If the required information is not available, more aggregate measures of the distributional effects of the policy will have to be used. One possibility would be to weight the aggregate changes in consumers' and producers' surpluses by the distribution characteristic for each good, as discussed in chapter 3.

The major difficulty in implementing the approach illustrated in figure 6-5 to obtain an estimate of the total change in consumers' and producers' surplus is the amount of information required for each commodity. In order to calculate area B_0FGB_1, it is necessary to know the locations and shapes of the original demand and supply curves and the shift in the supply curve that would result from a given change in air pollution. For most commodities, the available information on demand and supply curves will be limited to estimates of the price elasticities of demand and supply at existing prices and quantities. Also, only partial information will be available on the effects of changes in air pollution on supply curves. Although the change in total consumers' and producers' surplus cannot be calculated directly from such information, it can be approximated using the formulas discussed later in this chapter.

More-severe problems arise for those effects of air pollution experienced directly by individuals rather than indirectly through shifts in market supply curves. Important examples are the effects of air pollution on health and amenities. For most direct effects, the information required to calculate even the approximation formulas will not be available. Therefore, alternative procedures, which are discussed in chapters 8 and 11, will have to be used. Even in these cases, however, the methodology developed in this chapter can provide a useful conceptual framework for evaluating the estimates obtained. The basis for using the methodology in this way is the concept of household production.

The household-production approach views each household as using purchased commodities and time as inputs in the production of composite commodities from which utility is derived.[6] For example, households can be viewed as using food, medical care, exercise, and so forth, as inputs in the production of good health. Changes in environmental quality that affect health can then be thought of as shifting the household's supply curve for health by altering the quantities of health produced with given quantities of the other inputs. Therefore, the health benefits from improved environmental quality correspond to the increase in all households' consumers' and producers' surplus from health.

Reformulating the problem in this way makes it possible to evaluate potential errors in other procedures. For example, procedures that implicitly assume that household demand curves for health or amenities are vertical (that is, that no more will be desired when the cost is reduced) will underestimate the benefits of improved air quality. In some cases it may be possible to infer sufficient information on household demand and suppy curves to implement the approximation formulas that are derived for marketed commodities in the next section.

Approximation Formulas

The derivation of approximation formulas for the changes in total consumers' and producers' surplus resulting from an increase in air quality can be illustrated using figure 6-5. The change in total consumers' and producers' surplus (ΔCPS) is equal to the shaded area, B_0FGB_1. For the linear demand and supply curves in figure 6-5, the geometric formula for ΔCPS is

$$\Delta\text{CPS} = (B_0 - B_1) Q_0 + \tfrac{1}{2} (B_0 - B_1)(Q_1 - Q_0)$$
$$= \tfrac{1}{2} (B_0 - B_1)(Q_1 + Q_0) \qquad (6.1)$$

The formula for ΔCPS can be reexpressed in terms of prices, quantities, and elasticities of demand and supply. First, the equation for the original supply curve, S_0, is written

$$P_0^S = B_0 + \frac{\Delta P}{\Delta Q} Q_0 = B_0 + \frac{1}{\xi_0} \frac{P_0}{Q_0} Q_0 \qquad (6.2)$$

where P_0^S is the price required by suppliers, $\Delta P \equiv P_1 - P_0$ and $\Delta Q \equiv Q_1 - Q_0$ represent movements along the supply curve, and $\xi_0 \equiv (\Delta Q/\Delta P)(P_0/Q_0)$ is the elasticity of supply evaluated at P_0 and Q_0. If the change in air quality does not change the slope of the supply curve (that is, if it results in a parallel shift as in figure 6-5), then the equation for the new supply curve, S_1, can be written

$$P_1^S = B_1 + \frac{1}{\xi_0} \frac{P_0}{Q_0} Q_1 \qquad (6.3)$$

The equation for the demand curve can be written

$$P_0^D = A + \frac{\Delta P}{\Delta Q} Q_0 = A + \frac{1}{\eta_0} \frac{P_0}{Q_0} Q_0 \qquad (6.4)$$

where P_0^D is the price offered by purchasers, ΔP and ΔQ represent movements along the demand curve, and $\eta_0 \equiv (\Delta Q/\Delta P)(P_0/Q_0)$ is the elasticity of demand at P_0 and Q_0. The change in air quality results in a move along the demand curve from Q_0 to Q_1 and a decrease in the price offered by purchasers from P_0^D to P_1^D. Using equation 6.4, the value of P_1^D can be determined from the equation

$$P_1^D = P_0^D + \frac{1}{\eta_0} \frac{P_0}{Q_0} (Q_1 - Q_0) \qquad (6.5)$$

For markets to clear, the price required by suppliers must equal the price offered by purchasers, $P_1^S = P_1^D$. Therefore, from equations 6.3 and 6.5

Priced and Unpriced Commodities

$$B_1 + \frac{1}{\xi_0}\frac{P_0}{Q_0}Q_1 = P_0 + \frac{1}{\eta_0}\frac{P_0}{Q_0}(Q_I - Q_0) \tag{6.6}$$

Solving equation 6.6 for B_1, and equation 6.2 for B_0, and substituting in equation 6.1:

$$\Delta CPS = P_0 \Delta Q_1 \left(\frac{1}{\xi_0} - \frac{1}{\eta_0}\right)\left[1 + \frac{\Delta Q_1}{2Q_0}\right] \tag{6.7}$$

where $\Delta Q_1 = Q_1 - Q_0$.

Equations 6.1 and 6.7 both provide exact measures of the change in total consumers' and producers' surplus when demand and supply curves are linear. If demand and supply curves are not linear, then equations 6.1 and 6.7 can be interpreted as approximation formulas for ΔCPS.

Because equation 6.1 incorporates information on both the original and the new intercepts, B_0 and B_1, and quantities, Q_0 and Q_1, its use of linear approximations to the actual demand and supply curves generally will result in only small errors in estimating ΔCPS. However, the linear approximations underlying equation 6.7 incorporate information only on the original quantity, Q_0, and on the slopes of the demand and supply curves at that point. Therefore, the use of equation 6.7 to approximate ΔCPS may result in substantial errors. Unfortunately, the information on demand and supply curves required to implement equation 6.7 is far more likely to be available in practice than is the information required for equation 6.1. The use of qualitative information on supply curves to minimize the errors involved in using approximation formulas such as equation 6.7 is discussed later in this section.

In addition to information on demand and supply elasticities, implementation of equation 6.7 requires information on $Q_1 - Q_0$, which is the change in output of the affected commodity after all market adjustments are complete. Information of this type can be obtained from statistical analyses of data on output in localities with different amounts of air pollution.

Laboratory studies of the effects of air pollution generally provide information on the change in output that would occur if all other inputs were held constant, as illustrated by distance $Q_2 - Q_0$ in figure 6-5. Similarly, statistical studies of data on market prices or costs of production will yield information on the change in the cost per unit of the affected commodity after all market adjustments have occurred (distance $P_1 - P_0$ in figure 6-5), whereas laboratory studies on costs of production will yield data on the change in costs per unit, holding output constant (distance $P_2 - P_0$ in figure 6-5).[7]

Algebraic manipulation of equation 6.7 yields approximation formulas incorporating each of these alternative types of information on the effects of air pollution. To simplify notation, let $\Delta Q_2 = Q_2 - Q_0$, $\Delta P_1 = P_1 - P_0$, and $\Delta P_2 = P_2 - P_0$. The alternative forms of the approximation formulas are then as follows:[8]

$$\Delta CPS = P_0 \Delta Q_2 \frac{1}{\xi_0} \left[1 + \frac{\Delta Q_2}{2Q_0 \left(1 - \frac{\xi_0}{\eta_0}\right)} \right] \quad (6.8)$$

$$\Delta CPS = -Q_0 \Delta P_1 \left(1 - \frac{\eta_0}{\xi_0}\right) \left[1 + \frac{\eta_0 \Delta P_1}{2 P_0} \right] \quad (6.9)$$

$$\Delta CPS = -Q_0 \Delta P_2 \left[1 - \frac{\Delta P_2}{2 P_0 \left(\frac{1}{\xi_0} - \frac{1}{\eta_0}\right)} \right] \quad (6.10)$$

Most existing estimates of the values of pollution effects on commodities have been calculated using one of the following expressions:

$$P_0 \Delta Q_1, P_0 \Delta Q_2, -Q_0 \Delta P_1, -Q_0 \Delta P_2 \quad (6.11)$$

Since each of these expressions is equal to the first term in one of the approximation formulas, the accuracy of the existing estimates can be evaluated using the corresponding approximation formula.

If the change in quantity or price is small, the approximation formulas can be used to evaluate estimates such as those in 6.11 even if sufficient information is not available to calculate the approximation formulas directly. When the change in quantity or price is small, the term in square brackets in each formula will be approximately equal to 1.0, and the approximation formulas for ΔCPS become

$$\Delta CPS \approx P_0 \Delta Q_1 \left(\frac{1}{\xi_0} - \frac{1}{\eta_0}\right) \quad (6.7')$$

$$\Delta CPS \approx P_0 \Delta Q_2 \frac{1}{\xi_0} \quad (6.8')$$

$$\Delta CPS \approx -Q_0 \Delta P_1 \left(1 - \frac{\eta_0}{\xi_0}\right) \quad (6.9')$$

$$\Delta CPS \approx -Q_0 \Delta P_2 \quad (6.10')$$

Priced and Unpriced Commodities

Equation 6.7' indicates that $P_0 \Delta Q_1$ will provide a good approximation to ΔCPS if ΔQ_1 is small and $(1/\xi_0 - 1/\eta_0) = 1.0$ (for example, if $\xi_0 = |\eta_0| = 2.0$, where $|\eta_0|$ indicates the absolute value of η_0). The estimate $P_0 \Delta Q_1$ provides an underestimate if, as is more likely, $(1/\xi_0 - 1/\eta_0) > 1$ (for example, because ξ_0 and $|\eta_0|$ are both less than 2.0). Equation 6.8' indicates that $P_0 \Delta Q_2$ provides an appropriate estimate of ΔCPS if $\xi_0 = 1.0$, but an overestimate if $\xi_0 > 1.0$.[9] From equation 6.9' it can be seen that $-Q_0 \Delta P_1$ underestimates ΔCPS, because $(1 - \eta_0/\xi_0)$ is always greater than 1.0 (since $\eta_0 < 0$ and $\xi_0 > 0$). Finally, equation 6.10' shows that the expression $-Q_0 \Delta P_2$ provides an appropriate estimate of ΔCPS when the change in prices is small.

The approximation formulas discussed up to now have been based on the assumption that the supply curves have intercepts on the vertical axis, implying that no amount will be supplied at a price of 0, or, equivalently, that marginal cost is positive for all quantities greater than 0. It also has been assumed that the effect of a change in air quality is a horizontal shift in the supply curves, implying that the effect on cost per unit is a constant absolute amount. Formulas for the change in consumers' and producers' surplus for other sets of assumptions can be developed in an analogous way.

The appropriate formulas for two alternative sets of assumptions will be considered here: (1) the supply curves have intercepts on the horizontal axis,[10] and the effect of a change in air quality is a parallel shift in the supply curves; and (2) the supply curves are rays from the origin,[11] and the effect of a change in air quality is to rotate the supply curves, implying that the effect on costs per unit is a constant proportion. Case (1) is illustrated in figure 6-6(a) and case (2) is illustrated in figure 6-6(b).

Each of the approximation formulas can be written as the product of one of the expressions in 6.11 and a factor involving elasticities of demand and supply. For compactness, table 6-1 shows the multiplicative factors needed to calculate ΔCPS from each of these expressions. Part A of the table shows the general form of the multiplicative factors, and part B shows their approximate values when the change in quantity or price is small. The first column of the table summarizes the results discussed previously, the second column shows the formulas for the case illustrated in figure 6-6(a), and the third column shows the formulas for the case illustrated in figure 6-6(b).

The variation in the formulas across each row indicates the magnitude of the error that may occur when estimates of ΔCPS are based only on information on the elasticities of supply and demand. For example, the last row of part B shows that the estimate of ΔCPS provided by the formula in the first column is twice as large as the estimate provided by the formula in the last column. However, choice of the correct column requires only the determination of whether the supply curves intersect the vertical axis, intersect the horizontal axis, or are rays from the origin. Since this type of

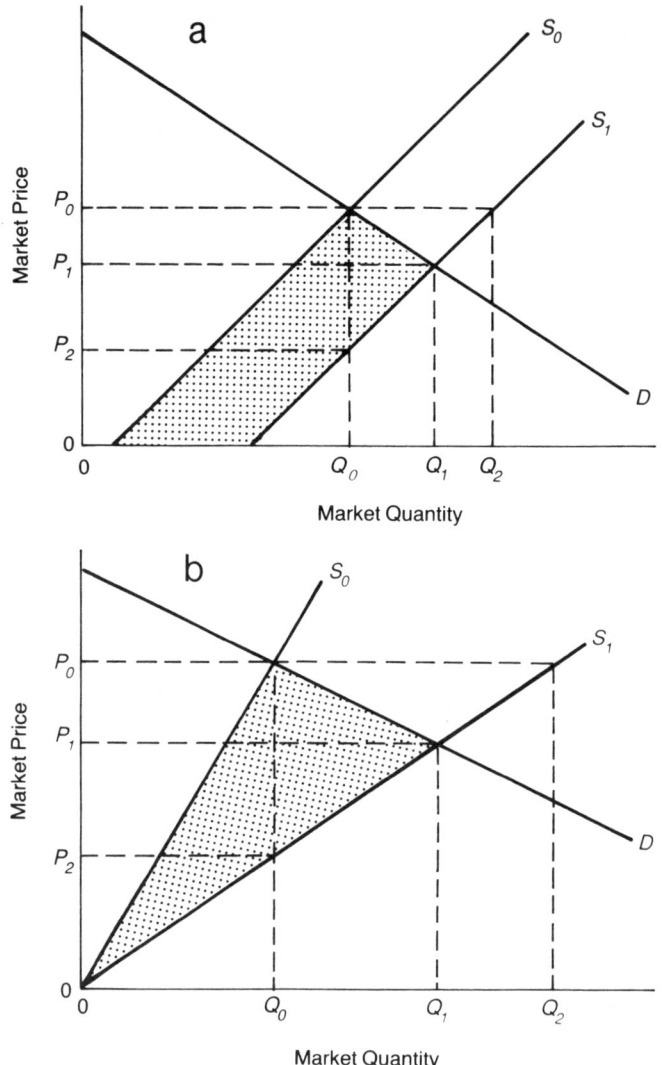

Figure 6-6. Alternative Supply Assumptions

qualitative information on supply curves often can be inferred from the characteristics of the commodity being considered, the major source of potential error involved in using the approximation formulas may be avoided. However, if one of the expressions in 6.11 is used to estimate ΔCPS, it will not be possible to avoid such errors by adjusting the estimate for differences in supply curves.

In some instances, estimates of the value of the effects of changes in air

Priced and Unpriced Commodities

quality have been based on the total expenditure on a commodity before ($E_0 = P_0Q_0$) and after ($E_1 = P_1Q_1$) a change in air quality. Approximation formulas based on these data can be derived in ways analogous to the derivations of the earlier formulas. For the case illustrated in figure 6-5, the formula is

$$\Delta \text{CPS} = \frac{1}{2}\left[E_0 - E_1 + E_0\left(\frac{Q_1}{Q_0}\right) - E_1\left(\frac{Q_0}{Q_1}\right)\right] \quad (6.12)$$

$$- \frac{1}{2\xi_0}\left[E_0 - E_0\left(\frac{Q_1}{Q_0}\right)^2\right]$$

Table 6-1
Multiplicative Factors for Estimating Total Consumers' and Producers' Surplus

First Term	Supply Curves Intersect the Vertical Axis	Supply Curves Intersect the Horizontal Axis	Supply Curves Are Rays from Origin
	A. Multiplicative Factors (ΔCPS ÷ First Term)		
$P_0 \Delta Q_1$	$\left(\frac{1}{\xi_0} - \frac{1}{\eta_0}\right)\left[1 + \frac{\Delta Q_1}{2Q_0}\right]$	$\left(1 - \frac{\xi_0}{\eta_0}\right)\left[1 + \frac{\Delta Q_1}{2\eta_0 Q_0}\right]$	$\frac{1}{2}\left(1 - \frac{1}{\eta_0}\right)$
$P_0 \Delta Q_2$	$\frac{1}{\xi_0}\left[1 + \frac{\Delta Q_2}{2Q_0\left(1 - \frac{\xi_0}{\eta_0}\right)}\right]$	$\left[1 + \frac{\Delta Q_2}{2Q_0\left(1 - \frac{\xi_0}{\eta_0}\right)}\right]$	$\frac{1}{2}\left[\frac{1 - \frac{1}{\eta_0}}{1 - \frac{Q_0 + \Delta Q_2}{\eta_0 Q_0}}\right]$
$-Q_0 \Delta P_1$	$\left(1 - \frac{\eta_0}{\xi_0}\right)\left[1 + \frac{\eta_0 \Delta P_1}{2P_0}\right]$	$(\xi_0 - \eta_0)\left[1 + \frac{\Delta P_1}{2P_0}\right]$	$\frac{1}{2}(1 - \eta_0)$
$-Q_0 \Delta P_2$	$\left[1 - \frac{\Delta P_2}{2P_0\left(\frac{1}{\xi_0} - \frac{1}{\eta_0}\right)}\right]$	$\xi_0\left[1 + \frac{\Delta P_2}{2P_0\left(\frac{\eta_0}{\xi_0} - 1\right)}\right]$	$\frac{1}{2}\left[\frac{1 - \frac{1}{\eta_0}}{1 - \frac{P_0 + \Delta P_2}{\eta_0 P_0}}\right]$
	B. Approximate Multiplicative Factors for Small Changes in Price or Quantity		
$P_0 \Delta Q_1$	$\frac{1}{\xi_0} - \frac{1}{\eta_0}$	$1 - \frac{\xi_0}{\eta_0}$	$\frac{1}{2}\left(1 - \frac{1}{\eta_0}\right)$
$P_0 \Delta Q_2$	$\frac{1}{\xi_0}$	1	$\frac{1}{2}$
$-Q_0 \Delta P_1$	$1 - \frac{\eta_0}{\xi_0}$	$\xi_0 - \eta_0$	$\frac{1}{2}(1 - \eta_0)$
$-Q_0 \Delta P_2$	1	ξ_0	1

For the case illustrated in figure 6-6(a) the formula is

$$\Delta CPS = \tfrac{1}{2}\left[E_0 - E_1 + E_1\left(\frac{Q_0}{Q_1}\right) - E_0\left(\frac{Q_1}{Q_0}\right)\right]\left(1 - \frac{\xi_0}{\eta_0}\right) \qquad (6.13)$$

and for the case illustrated in figure 6-6(b) the formula is

$$\Delta CPS = \tfrac{1}{2}\left[E_0\left(\frac{Q_1}{Q_0}\right) - E_1\left(\frac{Q_0}{Q_1}\right)\right] \qquad (6.14)$$

Studies using data on total expenditures have generally assumed that the change in total expenditure, $\Delta E = P_0Q_0 - P_1Q_1$, is an appropriate estimate of the value of the change in air quality. However, the formulas in table 6-1 show that the change in total expenditure provides useful information on the change in total consumers' and producers' surplus only in certain special cases.

For example, if the demand curve were horizontal (that is, infinitely elastic), the price of the commodity would not be affected by a change in air quality ($P_0 = P_1 = P_2$), so using the change in total expenditures as a measure of ΔCPS would be equivalent to the use of $P_0\Delta Q_1$ or $P_0\Delta Q_2$. From part B of table 6-1, if the change in quantity were small, ΔCPS would be approximately equal to $(1/\xi_0)\Delta E$ if supply curves had intercepts on the vertical axis, $\Delta CPS \approx \Delta E$ if they had intercepts on the horizontal axis, and $\Delta CPS \approx (1/\Delta E$ if the supply curves were rays from the origin. Similarly, if the demand curve were vertical (that is, infinitely inelastic), the quantity of the commodity would not be affected by a change in air pollution ($Q_0 = Q_1 = Q_2$), so that using the change in total expenditures would be equivalent to using $-Q_0\Delta P_1$ or $-Q\Delta P_2$. In this case, the relationship between ΔCPS and ΔE could be determined from the last row of part B of the table.

Summary

Market data provide useful information for estimating the value of a change in the quantity of a commodity consumed or produced. If markets are perfectly competitive, then the market price of a commodity will be equal to its marginal dollar value to consumers and to its marginal cost of production. Therefore, for small changes in the quantities consumed and produced, market prices provide the appropriate weights to use in aggregating over commodities. If the changes in the quantities consumed or produced are large enough to affect market prices, the appropriate measure of the dollar value of the changes is the change in total consumers' and producers' surplus. This can be approximated using an average of the prices before and after the change occurs.

Priced and Unpriced Commodities

If markets for the relevant commodities exist, but are not perfectly competitive, the prices paid by consumers will not be equal to the marginal costs of production. The change in total consumers' and producers' surplus remains an appropriate measure of the dollar value of the change in the quantities consumed and produced but cannot be as closely approximated using the average of before and after prices.

When a market for a relevant commodity, such as air quality, does not exist, market data for related commodities may be used to estimate the value of a change in the unpriced commodity. Approximation formulas have been developed for estimating the change in total consumers' and producers' surplus for the affected commodities. The accuracy of the approximations depends to a large extent on the characteristics of the supply curves. The types of estimates usually used do not permit these differences in supply characteristics to be taken into account.

Notes

1. Freeman (1979b) provides an excellent nontechnical discussion of the advantages and disadvantages of each of the principal measures that have been proposed as well as of Willig's resolution of the controversy.

2. The costs of supplying labor also will include any increases in the costs of transportation, clothing, food, and so on that are incurred by working, and other costs such as additional risks of accidental injury or death resulting from the job. Chapter 8 discusses procedures for using data on wage rates to infer the value of a reduction in the risk of accidental injuries or death.

3. As with consumers' surplus, alternative measures of the net benefits to a supplier of a price change have been proposed. The producers' surplus measure used here provides an adequate approximation to these other measures for most cases likely to arise in practice (Freeman 1979b).

4. If the price of the commodity were less than average variable cost, the firm would shut down.

5. McKean (1968) argues that the cost of trying to adjust market prices for market imperfections generally will be greater than the benefits of doing so. For opposing viewpoints, see the comments on McKean's paper by Kneese (1968) and Margolis (1968).

6. For a discussion of the household-production approach see Michael and Becker (1973).

7. If the demand curve is horizontal (that is, infinitely elastic), $Q_2 - Q_0$) will equal $(Q_1 - Q_0)$. If the demand curve is vertical (that is, infinitely inelastic), $(P_1 - P_0)$ will equal $(P_2 - P_0)$.

8. Equations 6.9 and 6.10 are multiplied by -1 because ΔP_1 and ΔP_2 are negative for increases in air quality.

9. For the case illustrated in figure 6–5, ξ_0 cannot be less than 1.0 because linear supply curves with intercepts on the vertical axis imply that $\xi_0 \geq 1$. Formulas for the case $\xi_0 \leq 1$ are developed later in this chapter.

10. For linear supply curves, this implies that $0 \leq \xi_0 \leq 1$.

11. For linear curves, this implies that the elasticity of supply is equal to 1, both before and after the change in air quality.

Part II
Application to Air-Pollution Control

7 Quantifying Air-Pollution Effects

An air-pollution-control policy may be stated either in terms of the maximum concentration of the pollutant that will be permitted in the air (often expressed as an *ambient standard*) or in terms of the amount of the pollutant that can be released by a source (often expressed as an *emission standard*). In the former case it is necessary to estimate the reduction in emissions required to avoid exceeding the ambient concentration limit, and in the latter case it is necessary to estimate the ambient concentration that will result from the specified reduction in emissions. In either instance, the expected change in concentration is used to estimate the effects on health, agricultural productivity, materials, and aesthetic values.

This chapter lays a foundation for later, specific discussions of benefit calculations. First, it outlines the use of mathematical air-quality models to predict either the required reductions in emissions (given an air quality) or the expected concentrations of pollutants (given a rate of emissions). Second, it describes how a change in air-pollution concentrations can be translated into a change in the quantity of various goods and services through *dose-response functions*. Finally, it discusses the statistical procedure of multivariate regression analysis, which is used in many studies of the relation between air-pollution concentrations and environmental damages. Appendix A provides a brief review of the effects of the more common air pollutants, as well as of their major sources, prevalence, and measurement techniques.

Estimating Pollutant Exposure

The concentration of pollutants often is expressed in terms of the total mass of the pollutant in a standard volume of air. In metric units the measure most frequently used is micrograms of pollutant in one cubic meter of air ($\mu g/m^3$). This measure can be used either for particles or for gases. Concentrations of gases also can be expressed as parts per million (ppm), where 1 ppm represents 1 cubic meter of the pollutant dispersed in 1 million cubic meters of air. A factor can be calculated for each pollutant gas to convert ppm into $\mu g/m^3$, or vice versa. For example, for sulfur dioxide at reference conditions, 1 ppm = $2,620 \mu g/m^3$.

The concentration of an air pollutant at a given point is a function of a number of variables, including the amount of the pollutant released at the source (the emission rate), the distance of the point from the source, and the atmospheric conditions. The most important atmospheric conditions are wind speed, wind direction, the amount of sunshine, and the vertical temperature characteristics of the local atmosphere. Most commonly the air temperature decreases with height, which results in an "unstable" atmosphere that tends to mix pollutants into the higher layer of the atmosphere, keeping pollution concentrations moderate or weak at ground level. If the temperature pattern is inverted in such a way that the upper air is warmer than the lower air, then the atmosphere will be "stable," with calm winds and potentially high pollution concentrations.

Dispersion Models

Various mathematical models have been developed to predict the concentration of pollutants at a given location, when the emissions and atmospheric conditions are known. Turner (1979) has published a review of many of these models and a discussion of many of the unresolved problems. The simplest model assumes that the area of interest can be approximated by a box, with the size of the base determined by local geography and the height fixed by atmospheric conditions. The emissions within the box and the winds that remove the pollutants from the box are assumed to be constant, resulting in a steady uniform concentration of pollution. The concentration can be calculated from

$$C = c + \frac{Q}{WHU} \qquad (7.1)$$

where C is the observed concentration; c is the background pollution entering the box; Q is the rate of emissions into the box (for example, kilograms per hour); W is the width of the box; H is the height of the box; and U is the wind speed. The limitations of such a simple model are apparent, but the box model can be helpful for quick approximations to define the magnitude of a problem.

A more realistic pollution model is the Gaussian plume-dispersion equation. In this model the concentration of pollution downwind from a source is assumed to spread outward from the centerline of the plume following a Gaussian (or normal) statistical distribution. The constants of the distribution are determined by the stability of the atmosphere and the "roughness" of the earth's surface in the vicinity. The plume spreads in both the horizontal and vertical directions, as illustrated in figure 7-1. The concentration at the surface is estimated by

Quantifying Air-Pollution Effects

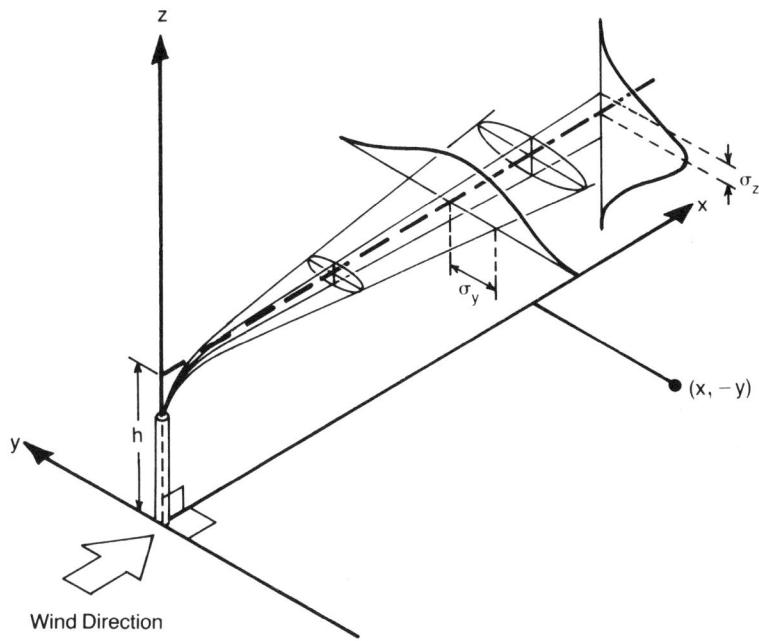

Figure 7-1. Gaussian Plume Model

$$C = c + \frac{Q}{2\pi \sigma_y \sigma_z U} \exp\left[- \tfrac{1}{2}\left(\frac{y^2}{\sigma_y^2} + \frac{h^2}{\sigma_z^2}\right)\right] \qquad (7.2)$$

where, in addition to the terms defined for equation 7.1, σ_y and σ_z are the Gaussian distribution parameters, y is the horizontal distance perpendicular to the wind direction, z is the vertical direction, and h is the effective height of the plume (considering the additional altitude to which the hot gases will rise above the height of the stack). Turner (1970) describes in detail the use of the Gaussian model and presents several tables and graphs that permit estimates to be made without going through the computations of equation 7.2.

This simple form of the Gaussian model does not consider changes in wind speed and direction at different altitudes or over time, variations in surface topography, or any other local influences on the meteorology (such as a body of water). It is possible to adjust the model to include some of these factors, but the meteorological data needed seldom are available. The accuracy of Gaussian models ranges from approximately ±20 percent under almost ideal conditions up to ±100 percent, depending on the distance from the source, the topography, and meteorological conditions (Systems Applications, Inc. 1975). In general, Gaussian models tend to underpredict near the source and overpredict away from the source.

Estimates for complex areas with many sources can be obtained by repeating the Gaussian model for each source and adding the results. Long-term averages of pollution concentrations can be estimated by calculating equation 7.2 for all wind directions and speeds and adding the results according to the probability that each combination will occur during an average year. Models also are available for "nonpoint" sources that stretch along a line, such as a highway, with its emissions of dust and automobile exhaust. Some models can estimate the concentrations that result from chemical reactions in the atmosphere or the natural removal of pollutants at the earth's surface. A few models have been developed that do not rely on the Gaussian approximation but, rather, numerically integrate the diffusion equation and the equations of motion and even take complex terrain into account. Substantial amounts of meteorological data are required to take full advantage of these more-accurate models. Both the simple and the complex models are available as computer programs. Several instruction manuals on the use of these programs have been published (for example, Khanna 1976; Fabrik, Skarlew, and Wilson 1977).

Ambient Data

Data from air-quality monitors may be used to obtain reasonable exposure estimates. Linear interpolations between values at two or three monitors are sometimes used to generate exposure estimates at the intermediate points. This can result in serious errors if there are significant topographical features between the monitors; if the data are not entirely commensurate (for example, the maximum twenty-four-hour average reading at two monitors may have occurred on different days under different meteorological conditions); if one or more of the monitors are dominated by local sources (such as a dusty parking lot); or if the concentrations are so low that the inherent monitor inaccuracy becomes a significant percentage of the value. The linear approximation may itself be a source of error. Benarie (1976) presents data showing that the expected concentrations often will decrease proportionally to the square of the distance from the source. His procedures also are useful for quick approximations of exposures resulting from emissions from a single point source, but they are not widely used.

When a monitoring network faces a single dominant source over relatively simply terrain, data from the monitors can be used in a simple "rollback" model to make a rough estimate of the change in the air pollution that can be expected if the emissions are changed. If the available monitoring data are due to a relatively constant level of pollutant emissions, then

$$C_1 = \frac{Q_1}{Q_0}(C_0 - c) + c \qquad (7.3)$$

where C_1 and Q_1 are the anticipated concentration and emission rate, respectively; C_0 and Q_0 are their initial values; and c is the background concentration. This equation also can be solved to estimate the reduction in emissions required to achieve a given concentration. More-accurate results can be obtained by correlating the air-pollution-monitoring data with the atmospheric characteristics. This technique has been described and demonstrated by Benarie et al. (1974).

Equation 7.3 cannot be applied to a pollutant, such as ozone, that results from an atmospheric chemical reaction. A similar, simple approximation permitting an estimate of ozone concentrations from the known concentrations of nitrogen oxides and hydrocarbons has been suggested by the U.S. Environmental Protection Agency (1977e). This procedure utilizes a graph that permits either an estimate of the change in the ozone concentration expected to result if changes are made in the emissions of the reacting pollutants, or an estimate of the amount of such changes necessary to achieve a given ozone concentration. A comparison of this method with other available techniques has been presented by DeMandel et al. (1979).

Population Distributions

It is also necessary to estimate the number and characteristics of the people, plants, materials, and so forth that are exposed to the pollutant at each location. The national census can provide the necessary population data in most instances. Projected changes in future U.S. population counts can be obtained from the OBERS estimates (U.S. Water Resources Council 1974) or from local public-utility agencies. The census reports also can be used to develop age- and income-specific population distributions. Farley (1978) has demonstrated the use of generally available data to generate distributions of pollution-sensitive population segments. Additional information on population-exposure analysis is given in U.S. Environmental Protection Agency (1977c).

Census figures refer to place of residence and therefore measure where people spend the nighttime hours. Since air-pollution patterns vary distinctly between nighttime and daytime, it would be useful to know the distribution of employment by census tracts. Information recovered from census data for business establishments generally is not satisfactory for this purpose, but this and similar information may be available from local planning agencies. The agencies that administer industrial insurance or unemployment compensation also may have records that can provide some locational information on employees, at least to the postal-code level.

It is difficult to obtain agricultural data for an area smaller than a county. It may be possible, using the county data as a guide, to develop more-specific distributions through interviews with county agricultural specialists or with farmers' or stockmans' organizations.

Distributions of materials that may be damaged by pollution have been generated almost entirely by applying national per capita averages to local populations. Although this has obvious limitations, it is at present the only approach realistically available. Maps showing existing structures are published for many major cities in the United States and are available from most city building departments or planning agencies. However, construction of an inventory from these maps is extremely tedious. The U.S. Environmental Protection Agency is now funding research to produce a methodology for generating reasonably accurate materials inventories with less effort.

Dose-Response Functions

The severity of reported effects generally increases as the exposure to air pollution increases. For example, at dilute concentrations of lead in the air, small amounts of lead can be measured in the blood of exposed persons, but with no apparent effect on health. At higher concentrations of lead in the air, increased amounts of lead and certain biochemical compounds are observed in the blood of some people; and reduced mental ability can be measured in some children. At still higher exposures, some people are diagnosed as anemic because of the lead. Finally, at very high concentrations of lead in the blood, actual nerve and brain damage are observed.

Also, an increasing percentage of the exposed population will exhibit any specific effect as the average exposure increases. This may be true because of variations in the actual exposure or because of variations in the ability of individuals to withstand (or adapt to) the effects of the pollutants. For example, persons who are already weakened by age, illness, poor nutrition, smoking, and so forth generally will be more susceptible to the adverse effects of air pollution than will the more healthy members of the population.

There is substantial evidence that the response to a pollutant often is related to the total dose of the pollutant, that is, the total quantity of the pollutant received over the time period of the exposure. The amount of response to an increasing dose of pollutant is described by a *dose-response function*. Dose-response functions can be reported either in terms of the magnitude of the effect on a single individual or in terms of the number of individuals (or percentage of the total population) exhibiting a given response at a particular dose.[1]

When an economic value can be estimated for the observed response, the dose-response function can be translated into terms of economic loss. This estimate as a function of dose is then called a *damage function*. All the damage functions for each of the different effects of a pollutant could be added to summarize the total economic gain or loss due to changes in the average concentration of the pollutant.

Forms of Dose-Response Functions

Three functional forms that frequently have been utilized in dose-response studies are illustrated in figure 7-2. The linear function in figure 7-2(a) assumes that there is a constant proportionality between the dose of the pollutant and the number of people (or leaves or paint samples) exhibiting a particular effect. At zero pollution there is a certain rate of occurrence of the effect unrelated to pollution. As the amount of pollution increases above zero there will be an increased incidence of the effect. If the pollutant has no effect at all at concentrations below a definite *threshold,* the dose-response function will appear as in figure 7-2(b). This is often called a *hockey-stick* function because of its shape. The *sigmoid* curve plotted in figure 7-2(c) is effectively a combination of the previous two forms. A relatively few sensitive individuals respond to a low dose of the pollutant. Above a less-well-defined threshold, the majority of the population is affected. At a high dose the relatively few, hardy individuals who have shown no effects at lower doses begin to respond. It is also possible for the sigmoid curve to be displaced to the right (as in figure 7-2(d)), showing a definite threshold even for sensitive individuals.

Experiments on the acute toxicity of pollutants to plants and animals frequently have produced dose-response functions of the sigmoid form. As a result, the sigmoid form often is assumed to be appropriate for the effects of pollution on human health. However, studies of the incidence of cancer and radiation-induced effects often have observed dose-response functions that are linear and without a threshold (Wilson 1978). Several studies of the human-health effects of air pollution have found that a linear form fits the data as well as the other forms tested (for example, Mendelsohn and Orcutt 1979). This may be in part a result of the limited range of observations available. If most of the observations are clustered over a narrow range, as illustrated in figure 7-2(d), the linear form may not be rejected statistically

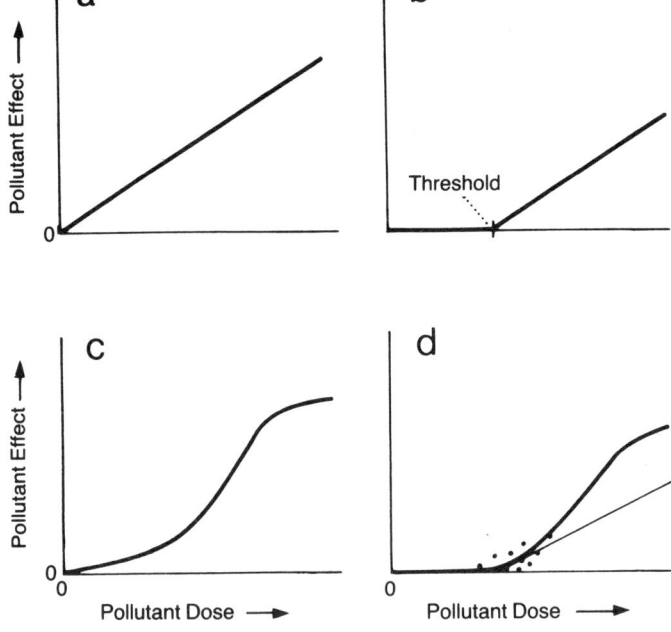

Figure 7-2. Dose-Response Functions

even if the true form of the dose-response function is actually sigmoid. In this case, predictions of effects obtained using the linear form would be adequate for small changes in pollution but could result in serious errors outside the range of the observations, again as illustrated in figure 7-2(d).

For low concentrations, one of the most important questions concerns the existence of a threshold of effects. If a threshold does exist, there can be no further reduction in effects when the pollutant dose is reduced below that point. Some studies of air-pollution effects have been interpreted as evidence for just such a threshold, but often it can be equally well argued that the apparent threshold is due to the design of the study or that other studies provide evidence for no threshold.

The initial intent of the ambient-air-quality standards adopted under the U.S. Clean Air Act was to reduce air pollution below such an absolute threshold. The law requires that the "primary" standard be set so as to "protect the public health" with "an adequate margin of safety." The "secondary" standard is intended to "protect the public welfare from any known or anticipated adverse effects." Therefore some benefit-cost studies do not attribute any economic loss to air pollution where the concentrations of pollutants are less than the U.S. standards. If the dose-response function for the pollutant did, in fact, exhibit a definite threshold and if the air-quality standard were established at that point, then the standard would indeed represent a threshold. Unfortunately, that is not always the case.

Although the U.S. law clearly excludes the consideration of any factors other than the scientific evidence of effects, the selection of an ambient-air-quality standard is a political decision subject to many pressures. Data indicating effects at lower concentrations might be given little weight so that a higher standard can be set, or a margin of safety might be added to reduce the standard below concentrations where adverse effects are reported. In either case the standard will fail to correspond to an actual threshold of effects.

Multivariate Regression Analysis

Studies of pollutant effects involve many more factors than the concentration of the pollutant and the time of the exposure. In general, air pollutants do not produce unique effects but, instead, aggravate or increase other effects that occur without any pollution present. Therefore, studies of effects must consider not just the changes in the pollutant dose received but also the changes in other variables that can alter the effects. Some of the most-useful studies are those of large and varied populations normally exposed to pollution in the ambient air. These are referred to as *field studies* in the study of plants and materials and as *epidemiologic studies* in the study of human-health effects. Clearly, such studies will involve analysis of a large number of potential influences on the effect in addition to the pollutant.

In order to separate the effects of the pollutant (or pollutants) from the effects of the other possible influences, analysts frequently use multivariate (or multiple) regression techniques. The first step is the identification of all the variables that may be relevant in explaining the changes in the effects. For example, if the effect of interest is death from respiratory disease, the relationship might be initially expressed as

$$M_{rd} = f(P_1, P_2, S, W, Z) \qquad (7.4)$$

where the "dependent variable," M_{rd}, is the mortality rate for respiratory disease; P_1 and P_2 are the observed average concentrations of two pollutants; S represents personal smoking habits; W is exposure to air contaminants in the workplace; and Z represents the list of other variables, such as nutrition, age, and sex, that are thought to affect the mortality rate.

The next step is the choice of a functional form for equation 7.4 that specifies the nature of the relationship between the mortality rate and the variables. For example, a linear functional form implies that mortality changes proportionally with the change in each variable. Using the linear form for illustrative purposes, and assuming that Z includes only one variable, the relationship to be estimated is

$$M_{rd} = a + bP_1 + cP_2 + dS + eW + fZ + u \qquad (7.5)$$

where the coefficients b, c, d, and so on indicate the effect of each variable on M_{rd}, and u is an error term that reflects all other sources of variation in the mortality rate.

Estimation of equation 7.5 requires data on mortality rates at different pollutant exposures. The data can be obtained by observing areas with different pollutant exposures at the same point in time (*cross-section data*) or by observing changes in pollutant exposures over time for a single area (*time-series data*). It is also possible to combine cross-section and time-series data (*pooled data*).

Once the data are at hand, the coefficients of equation 7.5 can be estimated using standard statistical procedures. The estimated coefficients provide a measure of the partial effect of each variable on the dependent variable. For example, the estimate of b in equation 7.5 is a measure of the effect on respiratory-disease mortality of changing the concentration of pollutant P_1 by one unit, holding all other variables constant. Therefore, the estimate of the coefficient b can be used to construct a dose-response function for this pollutant.

Sources of Error in Regression Analysis

The validity of the estimated coefficients in a multivariate regression depends on the appropriateness of the assumptions made at each step of the procedure and on the quality of the available data. This section discusses several of the more serious problems that can arise in practice.[2]

If a variable that has a significant effect on the dependent variable is omitted from the equation, the estimated coefficent of any included variable that is correlated with the omitted variable will be biased. For example, if the omitted variable is positively related to the dependent variable, the estimated coefficient of an included variable that is positively correlated with the omitted variable will be biased upward and the estimated coefficient of an included variable that is negatively correlated with the omitted variable will be biased downward.

A different type of problem arises when the included variables are highly correlated with each other (that is, multicollinear). Multicollinearity results in imprecise, but not biased, estimates of the coefficients. The estimates are imprecise in the sense that their estimated standard errors are relatively large, which may result in the estimated coefficients not being statistically significant. For example, if ozone and nitrate concentrations are highly correlated, inclusion of both of these variables in an equation for an agricultural crop might result in both having insignificant coefficients, even if both do reduce the yield of the crop. Omission of one of these variables would increase the apparent statistical significance of the other. However, since the included variable would now be correlated with a relevant omitted variable, the coefficient of the included variable would be biased. Thus it may be difficult to choose which variables to include in the equation. In such situations, it is desirable to have available the results of several alternative specifications of the equation.

In practice, the most-serious problem in obtaining reliable estimates of the effects of air pollutants through multiple regression analysis is the frequent lack of data of sufficient quality. In some cases a clearly relevant variable may have to be omitted from the equation because the data simply are not available and cannot reasonably be collected. As a result, the coefficients of the included variables may be biased, as discussed previously. In other cases it may be possible to include another variable that is believed to be highly correlated with the desired variable to serve as a surrogate or proxy. For example, "years of school completed" might be used as a surrogate for "nutritional habits." However, the data for a proxy variable cannot provide fully accurate measures of the variable it represents.

Even when direct measures of a variable are available, they may include errors that result in a biased estimate of its coefficient. For example, data from the Jacobs-Hochheiser method for measuring nitrogen dioxide have been used in many studies. Subsequently, it has been learned that the readings are frequently high. When data for a variable are systematically overstated, the estimated coefficient of the variable will be biased downward. Similarly, if the data for a variable are systematically understated, its estimated coefficient will be biased upward.

If the errors in a variable are not systematic but, instead, some values are too high and others are too low, the result will be a bias toward zero in the estimated coefficient of the variable. One example of this type of measurement error is the use of sulfation candles and plates as a measure of sulfur dioxide. Errors are created by wind speed and a number of other factors that result in readings that are often too high or too low by an unknown amount. Therefore, the use of these data will result in an underestimate of the effects of sulfur dioxide on the dependent variable.

The choice of an incorrect functional form for an equation will reduce the ability of the equation to explain variations in the dependent variable and, more important, may result in serious errors if the equation is used to predict the effect on the dependent variable of large changes in the explanatory variables. The first problem can be overcome by estimating a number of functional forms and choosing the one that "fits" the data best.[3] However, this procedure does not guarantee that the equations will yield reliable predictions for values of the variables lying outside the observed range.

Summary

In order to quantify the effects of an air-pollution-control policy, it is necessary to know the change in the pollutant exposures that can be expected to occur and to have dose-response functions that allow the calculation of the magnitude of the effects of the changed exposures. Air-quality models or existing monitoring data can be used to estimate the change in pollution exposures. Dose-response functions can be constructed using multiple regression analysis, but the analyst should be aware of the sources of potential error in the estimated coefficients.

Notes

1. An earlier review of dose-response functions by Hershaft, Morton, and Shea (1976) may be consulted for additional discussion.
2. Any standard econometrics textbook (for example, Johnston 1972) may be consulted for further discussion of these topics.
3. See Halvorsen and Pollakowski (1981) for a discussion of a general statistical procedure for the choice of functional forms.

8 Estimating Health Benefits

The effects of air pollution on human health range from slight increases of pollutants in the body (body burden), through physiologic changes of uncertain consequence, to the clearly pathologic changes of illness and death. Health effects commonly associated with air pollution include eye irritation, tightness in the chest, scratchy throat, aggravation of asthma, and increased respiratory illness. Air pollutants also have been associated with reduced tolerance for exercise, reduced mental ability, nausea, anemia, heart disease, kidney damage, leukemia, cancer, and other ailments.

Estimating the health benefits from air-pollution control requires information both on the dose-response function for each effect and each pollutant and on the economic value of each effect. This information then can be combined into a damage function that gives the dollar value of the health effects for each pollutant. The benefit of reducing air pollution is then measured by the resulting reduction in damages.

This chapter discusses the principal sources of information on damage functions for health. Two types of dose-response studies are discussed: statistical analyses and subjective assessments. Studies that provide estimates of the economic value of the effects are also described. These include studies of wage rates in hazardous industries that can be used to infer the dollar value of a reduction in risks to life and health.

Development of Health-Effects Dose-Response Functions

Laboratory studies of pollutant effects have tended to focus on the degree of impairment observed in the average individual studied at a given pollutant concentration. These studies have been concerned primarily with body burden and small physiological changes and often define the response observed in a way that makes it difficult to compare the results with observations in other studies. The measurements often have been made at only two or three concentrations.

Few useful dose-response functions can be derived from the extensive literature of laboratory studies. Babcock and Nagda (1976) used the data in twenty-three such studies to construct dose-response functions for reductions in lung function by exposures to ozone, sulfur dioxide, and nitrogen dioxide; but the ultimate effect on health of these immediate lung function changes is not well understood. Although repeated exposures may have long-term consequences for health, measurements of such effects are not available. Without that information it is difficult to determine what value might be placed on the effects. In addition, laboratory studies often restrict their choice of subjects to healthy young individuals, so the data can not provide an overall estimate of the effects on the general population.

Studies of carbon monoxide, lead, and some other pollutants have produced reliable dose-response functions in terms of the concentration of the pollutant in the blood, as measured by various laboratory tests. For several of these pollutants, equations have been developed that relate the ambient exposure (both concentration and time) to the blood concentrations. However, the resulting relation between concentrations in the air and response is complex, and the assumptions necessary to make the often tedious calculations limit the usefulness of the derived functions. For other pollutants, blood concentration or other simple measurements may not accurately characterize the concentrations in critical body organs; or the relation of atmospheric concentrations to body burden may be poorly understood. Walsh, Killough, and Rohwer (1978) have developed a specific procedure for computing a dose-response function for pollutants that accumulate in the body.

Epidemiologic studies of the relationship between health status and air pollution are more useful in developing dose-response functions. These studies use measures of the health status of individuals exposed to actual, normally occurring air pollution. By taking advantage of such "natural experiments," an epidemiologic study is able to gather information on the effects of chronic exposures and on more varied groups of individuals (although the response of the most vulnerable segments of the population remain difficult to observe since these individuals often are receiving medication to relieve the response). As suggested by Shy (1979), these studies may be discussed as two distinct types, micro- and macroepidemiologic studies.

A microepidemiologic study observes the health status of specific, identified individuals. Therefore, it is possible to collect detailed information on the socioeconomic characteristics, personal habits, and health history of the study population. Although this information permits a careful evaluation of the effects of the pollutant, the expense of such studies limits their scope. The results generally have been confined to statistical tests of whether the difference in the responses observed is due simply to chance. In most instances there is not sufficient variation within the study to permit construction of a dose-response function. However, evaluation of a number of such studies with a range of exposures and groups of subjects may suggest a dose-response function or functions. Dose-response functions developed by such subjective assessment will be discussed later in this chapter.

Estimating Health Benefits

Rather than observing specific individuals, a macroepidemiologic study uses data from large geographic units. The use of aggregate data makes it possible to use a substantial amount of information collected for other purposes. This permits more variables to be considered in the analysis. Most often, multiple regression techniques are used to separate the health effects that can be attributed to pollution from the effects of other differences in the population. The estimated equations are then used directly as dose-response functions.

The health effects measured in a macroepidemiologic study reflect the ongoing adjustments of the population to the pollution and, therefore, correspond to the distance $Q_1 - Q_0$ in figure 6-5.[1] The microepidemiologic studies and laboratory studies are based on closed panels of subjects, so the health effects they measure will indicate the change that will occur without adjustment, or distance $Q_2 - Q_0$ in figure 6-5.

Macroepidemiologic Studies of Mortality and Morbidity

Macroepidemiologic studies have been a major source of health-effects dose-response functions in benefit-cost analyses. Most have attempted to develop relationships between air-quality data and death rates (mortality). A few have attempted to develop the relationships to measures of illness (morbidity), such as hospital admissions. The better-known studies, their results, and some of the problems associated with such studies are reviewed briefly here. Detailed reviews of these and similar studies have been published by Herman (1977), Spengler et al. (1979), and Ricci and Wyzga (1979, 1981).

Mortality Studies

Seven cross-sectional and three time-series macroepidemiologic mortality studies are summarized in table 8-1. For each study the table presents the number and type of geographic units that form the basis of the study; the date of the data used; the socioeconomic and air-pollution variables included in the regression equations; and the values of the regression coefficients for total suspended particulate (TSP), sulfur dioxide, and sulfate, if included in the analysis.

The coefficients in table 8-1 have been converted to common units to permit comparison among the studies of the estimated marginal effects of each pollutant. However, the coefficients do not indicate the relative effect of the pollutants on total mortality since the average values and the variability (standard deviations) of TSP and sulfur dioxide are much larger than

Table 8-1
Macroepidemiologic Studies of Air Pollution and Mortality in the United States

Study	Data Base	Pollutant Coefficients				Other Variables Included	
		TSP	Sulfur Dioxide	Sulfate	Pollutants	Socioeconomic	Other
Cross-sectional							
Lave and Seskin (1977)	117 SMSAs: 1960	0.04*	X	0.4*	X	Age, race, density, SMSA population, poverty.	
	117 SMSAs: 1960	0.04*	X	0.2*	X	As the preceding plus occupation, fuels, or climate.	
	112 SMSAs: 1969	0.06*	X	0.4*	X	Age, race, density, SMSA population, poverty.	
	69 SMSAs: 1969	0.06*	0.05*	0.2*	NO_2^*, NO_3	As the preceding.	
Schwing and McDonald (1976)	46 SMSAs: 1960 (pollution data: 1965)	X	a	0.06*	NO_3, HC^a	(White male only) age, race, poverty, density, housing, education, SMSA population, family size, occupation, temperature, humidity, precipitation, sunshine, background radiation, smoking	
Liu and Yu (1976)	40 SMSAs: 1968–1970	0.01	0.02*	X	X	Age, race, education, poverty humidity, sunshine, thunderstorms.	
Gregor (1977)	Allegheny County, Penn.: 1968–1972	0.2*	0.02	X	X	(White only) education, density, precipitation, temperature.	

Estimating Health Benefits

Study	Area/Data	TSP	SO$_2$	Sulfates (0.6*)	Other pollutants	Variables controlled
Mendelsohn and Orcutt (1979)	404 county groups (entire U.S.): 1970 pollution data: 1974	−0.0			CO*, O$_3$, NO$_2$, NO$_3$	(White only) age, education, income, marital status, family size, density, housing, temperature, humidity, geographic regions.
Crocker et al. (1979)	60 cities: 1970	0.01	0.07	X	NO$_2$	Age, race, housing, temperature, doctors, smoking, diet.
Lipfert (1980)	181 cities 1969–1971	0.08*	−0.03	b	Mn*, Fe	Age, race, birth rate, poverty, housing, density, smoking.
Time Series						
Lave and Seskin (1977)	Chicago, Ill.: 1962–1964	X	0.03*	X	X	Day of the week, wind, temperature, rain.
Schimmel and Murawski (1976)	New York, N.Y.: 1963–1972	0.07*^{d}–0.16	0.01–0.04	X	X	Temperature, two-week trends.
Wyzga (1978)	Philadelphia, Penn.: 1957–1966	0.02*^{d}–0.04	X	X	X	Flu epidemics, season.

Notes: All regression coefficients have been converted to incremental annual deaths per 10,000 population per $\mu g/m^3$ change in pollutant (annual average).

SMSA = Standard metropolitan statistical area.

* = Statistically significant at the 95-percent confidence level or above.

X = Not included in regression.

^{a}Surrogate variable used that cannot be expressed in conventional units.

^{b}Included in regression but value not reported in study because of low statistical significance.

^{c}Gregor interpreted his data base for sulfur dioxide as in "parts per trillion." However, a reexamination of his data base reveals that the correct units are "parts per billion" (Anthony J. Sadar, Allegheny County Health Department, Bureau of Air Pollution Control, personal communication, 1980). Gregor's reported coefficients have been adjusted by a factor of 10^3.

^{d}Converted from COH with the factor 1 COH = 150 $\mu g/m^3$ of TSP.

those for sulfate. Adjusted for this, these studies present a consistent picture of TSP as the most-important air-quality variable, with sulfate second, and sulfur dioxide third.

The studies by Lave and Seskin (1977) are perhaps the best known and most thoroughly reviewed (see, for example, Thibodeau et al. 1980) of those listed in table 8-1. They reported coefficients for more than 400 regression equations that use a variety of data for mortality, air quality, and socioeconomic characteristics. In some equations additional variables were included for occupations, home-heating fuels, climate, and migration patterns. Most of the estimated equations assume a linear form, but other functional forms were tested.

Lave and Seskin used the annual minimum, mean, and maximum values of TSP and sulfate as the air-pollution variables in many of their equations. Because this makes comparison of their coefficients difficult, they reported a total elasticity for each pollutant in each equation. The linear coefficients reported here are computed from averages of the elasticities, evaluated at the mean values of the pollutant and mortality. The coefficients for TSP are consistent throughout most of their alternative equations. The coefficients for sulfate tended to be reduced when additional pollutants or socioeconomic variables were added to the equation.

Schwing and McDonald (1976) examined the effects of multicollinearity in their variables on the estimated coefficients using ridge regression techniques. The ridge regression method tests the robustness of the coefficients by introducing an artificial distortion into the variables. This procedure reduced the estimated coefficient of sulfate reported in table 8-1 by about one-third and the estimated coefficient of nitrate by two orders of magnitude.

Liu and Yu (1976) conducted a two-stage regression in order to utilize a sigmoid form for the air-pollution variables. In the first step they constructed a linear equation in the socioeconomic and climate variables. The remaining variation in the mortality rates not accounted for by this equation was then used in the sigmoid-form regression with the air-pollution variables. To the extent that the socioeconomic variables are positively correlated with the air-pollution variables, this two-stage procedure will lead to underestimates of the pollutant coefficients. The coefficients they obtained resulted in an equation that is approximately linear over the range of interest ($35 \mu g/m^3 < SO_2 < 120 \mu g/m^3$). The coefficients reported in table 8-1 are the linear equivalents of their coefficients for this range.

One of the more ambitious studies of mortality rates and air quality is the analysis of Mendelsohn and Orcutt (1979). Their data base included more than 2 million 1970 death certificates catalogued according to residence, age, sex, race, and underlying cause of death. The use of these data moves in the direction of a microepidemiologic study. However, other socioeconomic characteristics and air-quality variables were represented by aggregate data for one or more counties. Their regression coefficients are reported for the white population by classes specific to age, sex, and cause of death. The average coefficients reported in table 8-1 are calculated from these data (Mendelsohn and Orcutt 1977) and from the 1970 age and sex distribution of the U.S. population.

Variables representing acknowledged causes of poor health, such as diet, cigarette use, and the availability of medical care, have not been included in most macroepidemiologic studies. However, Crocker et al. (1979) did include variables on consumption of protein, fats, and carbohydrates; per capita sales of cigarettes; and the number of doctors per capita. They found that as air pollution increased, the availability of doctors decreased. They suggested that one source of the association between air pollution and mortality found by other analysts may be the preference of doctors not to live in polluted areas.

Because the three time-series studies listed in table 8-1 are based on daily air-pollution measurements and daily variations in mortality, they are expected to reflect acute responses to high, short-term exposures. Each study included variables for pollution concentrations up to five days prior to the date of the mortality data. The strongest association was for the same day, with the total effect approximately doubling when all days were included.

Lave and Seskin (1977) conducted time-series analyses for five U.S. cities for sulfur dioxide, carbon monoxide, hydrocarbons, and nitrogen oxides. They found statistically significant, positive coefficients only for the association with sulfur dioxide in Chicago. Lave and Seskin suggested that this may have been true because the number of daily deaths in Chicago was far larger than in the other cities they examined and because the average concentrations of sulfur dioxide were four to ten times higher in Chicago than in the other cities.

The Lave and Seskin analysis did not include a variable for particulate matter. Schimmel and Murawski (1976) analyzed daily mortality in New York City with respect to the coefficient-of-haze (COH) measure of particulate matter as well as sulfur dioxide. Wyzga (1978) conducted a similar analysis of daily mortality in Philadelphia, using COH alone as the pollution variable. In order to compare their estimated coefficients to those of other analysts, it is necessary to convert from COH to an equivalent measure of TSP in $\mu g/m^3$. This was done for the data reported in table 8-1.[2]

Morbidity Studies

Data on the incidence of illness are not regularly collected on a national basis by any agency. Therefore, macroepidemiological studies of the relationship between air pollution and the incidence of specific diseases have had to rely either on information collected specifically for the study or on data only marginally related to the health effect being observed.

Fishelson and Graves (1978) and a few other analysts have used hospital-emergency-room admissions as a source of data for time-series analyses of air-pollution morbidity effects. The study by Fishelson and Graves found a significant relation between cardiac admissions and the previous day's sulfur-dioxide concentration of 5×10^{-4} admissions per 10,000 population per $\mu g/m^3$ (twenty-four hour average). If a reliable estimate could be made of the fraction of aggravated heart-disease symptoms that result in an emergency-room visit, the total impact on heart-disease morbidity of a change in pollution levels could be estimated.

Regression techniques also have been applied to aggregated data from health-status surveys. One group of surveys that has been used by several analysts is the U.S. Environmental Protection Agency's Community Health and Environmental Surveillance System (CHESS) studies on respiratory-disease prevalence (U.S. Environmental Protection Agency 1974b). Both the health data and the air-quality data from these studies have been severely criticized (U.S. House of Representatives 1976b) as containing numerous potential sources of error. An example of the use of these data is the study by Leaderer, Berman, and Stolwijk (1977). They pooled the CHESS data with a similar data set of their own and regressed the combined incidence data against sex, smoking-habits, sulfur-dioxide, TSP, and sulfate-pollution data. The sulfur-dioxide and TSP coefficients were small and not statistically significant. The sulfate coefficient was 62 annual incremental cases of chronic bronchitis per 10,000 population per $\mu g/m^3$ change (annual average) with an apparent threshold of effects at $5.8 \mu g/m^3$.

Crocker et al. (1979) utilized data on absence from work as a measure of morbidity. The data were taken from a nine-year University of Michigan study of approximately 5,000 families from across the United States. Air-pollution data were available from monitors in 118 counties, which permitted assignment of air-pollution values to about 3,000 families. The large amount of data allowed Crocker et al. to include many socioeconomic variables. Four recursive equations for illness, hourly earnings, and annual hours of work were estimated. The results of the regressions are similar from year to year but are not entirely consistent, partly because the same socioeconomic data were not collected each year. The equations imply that a reduction in air pollution would reduce illness and increase both the annual hours of work and hourly wage rates. The approach taken by this study goes beyond the estimation of dose-response functions to a measurement of the effect of air pollution on household production and consumption. Although the results of this study must be treated as only indicative, the approach promises to be of significant future importance to the benefit-cost analysis of air-pollution control.

Sources of Error in Macroepidemiologic Studies

If variables that adversely affect health are omitted from the regression equation, the estimated coefficient of an air-pollution variable may not accurately reflect the effect of air pollution on health. If the omitted variables are positively correlated with the air-pollution variables, then the estimated effects of air pollution on health will be overstated. For example, observed concentrations of some air pollutants tend to be higher in the older, industrial cities of the northeastern United States. Therefore, other adverse influences on health that may be greater in these cities (such as noise, stress, crime, and so on) or associated with the European ethnic minorities concentrated in these cities (such as diet) might represent important omitted variables that are positively correlated with the air-pollution variables.

Several variables that might be assumed to be important determinants of health often either have not been included or have been represented by proxy variables because data are not available for the geographic units used in the macroepidemiologic studies. For example, smoking habits have been represented in some studies by data on state-tobacco-tax collections. These data will only approximately represent smoking habits. Crocker et al. (1979) constructed diet variables from diet-survey information aggregated by income classes for the four U.S. census regions. They then generated the values for each city from the regional diet data and the income distribution in that city. The use of these data produced implausible results (for example, that the mortality rate increases with decreased consumption of animal fats).

Occupational exposures to pollutants have been accounted for in a few studies by census data on the percentage of the work force in various industries or occupations. However, the industrial sectors reported are so broad that the data may be a better index of pollutant emissions than of occupational exposures. Exposures to indoor pollutants away from the workplace have not been included in any of the macroepidemiologic studies.

An indirect estimate of the importance of the biases from omitting smoking and occupational variables can be obtained from the sex-specific regression results of several analysts. The incidence of both occupational exposure and smoking is higher for men than for women. There are approximately 1.4 times as many smokers among men, and approximately five times as many men work in nonclerical blue-collar occupations. Therefore, such biases, if they exist, should be greater in the regressions for men than in those for women.

The results in the three studies that report sex-specific equations indicate that biases resulting from the omission of occupational and smoking variables may exist but probably are not the primary source of the apparent relation between air pollution and health. The estimated coefficients for particulate matter reported by Gregor (1977) and Lave and Seskin (1977) average approximately 20 percent greater for men than women. The estimated coefficients for sulfur dioxide and sulfates reported by Gregor (1977) and Mendelsohn and Orcutt (1977) show the same pattern, but the estimates reported by Lave and Seskin for sulfur dioxide and sulfates are larger for women than for men.

The poor quality of much of the air-quality data also may result in biases in the estimated coefficients of the air-pollution variables. For many years, most cities had no more than one air-quality-monitoring station. Because the monitor was almost invariably located in the more-polluted central-city area, the readings from these stations will tend to overstate the exposures of the community. This will bias the estimated coefficients of the air-pollution variables toward zero. Sulfates and ozone tend to be relatively uniform over large areas, so less bias from central-city measurements would be expected in the estimated coefficients of these pollutants.

Random errors in the air-quality data may bias the coefficients of the air-pollution variables toward zero. For example, the use of data on total suspended particulates (TSP) may result in an underestimate of the health effects of particulate matter since only a fraction of the total TSP measurement (the fine, or respirable, particles) is assumed to be related to adverse health effects. TSP also includes the coarse particles (often due to windblown dust), which vary randomly with respect to the fine fraction.[3] Random errors also may result from the failure of a single monitor adequately to represent the population exposure, even where central-city bias is not a problem (Goldstein and Landovitz 1977).

Because the existing air-quality data are not perfectly representative of population exposures, some macroepidemiologic studies have relied on pollutant-emissions data or have constructed surrogate variables from pollutant-emissions and local meteorological data. Schwing and McDonald (1976), for example, used the "pollution-potential" estimates of Benedict, Miller, and Smith (1973), which are described in chapter 9. These estimates appear to be only poorly related to actual air-quality measurements. In addition, the data used by Schwing and McDonald were from an early draft of the Benedict, Miller, and Smith report and were revised substantially in the final report.

Conclusions

The slope of the mortality dose-response curves for total suspended particulate, sulfur dioxide, and sulfates is suggested by the studies listed in table 8-1. However, a number of cautions must be raised. The two regressions by Lave and Seskin (1977) that include additional variables have lower estimated sulfate coefficients, suggesting that the estimated coefficients in the other equations may be increased by omitted variable bias. The two-stage procedure used by Liu and Yu (1976) may have biased their estimated coefficients toward zero. The use of sulfation data quite probably biased toward zero the sulfur-dioxide coefficients estimated by Gregor (1977). The near-zero value for Mendelsohn and Orcutt's (1979) estimated TSP coefficient may be the result of bias from random errors from the large number of rural monitors with a high proportion of coarse particulate in the western counties. As urban TSP and sulfate have been found by others to be moderately correlated, this may have resulted in an upward bias in their estimated sulfate coefficient. Each of these studies failed to include variables for smoking habits, occupational exposure, and diet, which may result in an upward bias in their estimated coefficients. The use by Crocker et al. (1979) of inappropriate diet variables, which are moderately correlated with the pollution variables, may account for the unusual coefficients they obtained. The conversion of the estimated-particulate-matter coefficients in the time-series studies from COH to TSP units increases the uncertainty there.

Microepidemiologic evidence for increased mortality from elevated concentrations of total suspended particulate matter and sulfur dioxide is almost entirely at ambient concentrations higher than those appropriate to most contemporary benefit-cost analyses in the United States, although increased morbidity is observed at the lower concentrations. The evidence for increased mortality or morbidity from elevated concentrations of sulfates is much more equivocal. Although some epidemiologic studies have identified sulfates as a significant species in explaining their data, laboratory studies on human subjects generally have failed to observe any substantial effects at moderate concentrations.

Based on these considerations, an evaluation of the data in table 8-1 suggests approximately 0.05 incremental annual deaths per 10,000 population per $\mu g/m^3$ change in the annual average of TSP and 0.04 incremental annual deaths per 10,000 population per $\mu g/m^3$ change in the annual average of sulfur dioxide.

The studies summarized in table 8-1 also suggest approximately 0.3 incremental deaths per 10,000 population per $\mu g/m^3$ change in the annual average sulfate concentration. However, there is substantial uncertainty as to whether a causal relation between sulfate exposure and mortality actually exists.

The question of the existence or location of effects thresholds for these pollutants is also highly controversial. Some studies have suggested that no thresholds exist, whereas others have suggested definite thresholds, often above the concentrations established as ambient-air-quality standards in the United States.

Subjective Assessment of Dose-Response Functions

An alternative approach to macroepidemiologic studies is the subjective definition of a dose-response function by one or more experienced researchers, based on their own scientific understanding of the pollutant and its effects. Although a consistency of scientific opinion (perhaps simply based on reading all the same reports) is not an adequate substitute for objective facts, it may provide useful information for policy analysis.

Two approaches have been taken in establishing subjective dose-response functions. In one, a small group of researchers gather data from laboratory and microepidemiologic studies and apply regression analysis, professional judgment, and personal opinions to the data to estimate a dose-response function. In the second approach, a larger number of researchers is interviewed to determine what each believes to be the effects associated with various doses of the pollutant. Their responses are then analyzed to obtain an estimate of the dose-response function.

An example of the first approach is the ambitious effort by Nelson, Knelson, and Hasselblad (1976) to estimate damage functions for four pollutants for death and thirteen other health effects. They assumed a linear dose-response function with a threshold. Some of their dose-response functions are based on studies that have been criticized sharply by some scientists. The sulfate functions, in particular, are open to question because they are based almost exclusively on the U.S. Environmental Protection Agency's CHESS studies.

Their reported functions are specific to the more-vulnerable age groups and populations (such as children, the elderly, persons with asthma, and so forth), but are presented in table 8-2 in simplified form for an average U.S. population. For the functions expressed in terms of an annual average, the number of excess events is $\alpha(C - \beta)$, where α is the coefficient listed in table 8-2, C is the annual average pollutant concentration, and β is the threshold.

For the functions expressed in terms of daily or hourly average concentrations, it is necessary to add up each hour or day during the year when the concentration will be at a value above the threshold and to multiply that count by the concentration and the dose-response coefficient. If the ambient concentrations throughout the year can be represented by a statistical distribution, the calculations are simplified considerably. For example, if a log-normal distribution is assumed, as is appropriate for most pollutants, then the number of excess events over N periods can be calculated as

$$\frac{N\alpha}{\sqrt{2\pi} \ln \sigma_g} \int_\beta^\infty (C - \beta) \exp\left[-\frac{\ln^2 (C/\bar{C})}{2 \ln^2 \sigma_g}\right] dC \quad (8.1)$$

Table 8-2
Nelson, Knelson, and Hasselblad Dose-Response Functions

	Threshold	Averaging Time	Coefficient
Carbon monoxide			
Mortality	13 mg/m³	8 hours	3.4×10^{-3}
Angina attack	9 mg/m³	8 hours	1.2
Ozone			
Aggravation of heart and lung symptoms	400 μg/m³	1 hour	5.8×10^{-4}
Aggravation of asthma	400 μg/m³	1 hour	5.8×10^{-5}
Eye discomfort	260 μg/m³	1 hour	8.5×10^{-3}
Cough	400 μg/m³	1 hour	9.1×10^{-3}
Chest discomfort	420 μg/m³	1 hour	1.0×10^{-3}
Headache	100 μg/m³	1 hour	1.8×10^{-3}
Nitrogen Dioxide			
Juvenile lower-respiratory disease	50 μg/m³	Annual	2.4
Sulfates			
Mortality	25 μg/m³	24 hours	6.5×10^{-2}
Aggravation of heart and lung symptoms	9 μg/m³	24 hours	0.84
Aggravation of asthma	6 μg/m³	24 hours	0.20
Juvenile lower-respiratory disease	13 μg/m³	Annual	90.0
Chronic bronchitis	12 μg/m³	Annual	34.0

Source: Adapted from William C. Nelson, John H. Knelson, and Victor Hasselblad, Air pollution health effects estimation model, in *Proceedings of the Conference on Environmental Modeling and Simulation,* EPA 600/9-76-016, pp. 191-195 (U.S. Environmental Protection Agency, 1976), by combining information presented in their paper to represent a national average population. Carbon-monoxide function converted to concentration units using data in J.E. Peterson and R.D. Stewart, Predicting the carboxyhemoglobin levels resulting from carbon monoxide exposures, *Journal of Applied Physiology* 39 (1975):633-638. This conversion assumes that the subject is continuously exposed to a constant carbon-monoxide concentration for the entire period.

Note: Units for carbon-monoxide coefficient are incremental events hourly per 10,000 population per mg/m³ increase above threshold as measured by an eight-hour average ending that hour. Units for all other pollutants are incremental events or cases per 10,000 population per unit of averaging time per μg/m³ of pollutant increase above threshold.

where $\bar{C}$ is the annual average pollution concentration and σ_g is the standard deviation of the distribution. A close approximation can be obtained in a numerical integration to an upper limit about eight times $\bar{C}$. Numerical tabulations of the annual excess events for the ozone, carbon-monoxide, and nitrogen-dioxide dose-response functions have been published (U.S. Interagency Task Force (1976) for various values of $\bar{C}$ and σ_g. Application of these functions is illustrated in a case study by North and Merkhofer (1975). The U.S. Council on Environmental Quality (1978) has used these functions to estimate the health risks of automobile-related pollutants in nine U.S. cities.

The second approach to the subjective construction of dose-response functions integrates the opinions of a number of experienced researchers. Leung, Goldstein, and Dalkey (1978) interviewed fourteen scientists separately, asking for their estimates of the concentrations of a pollutant at which 10 percent, 50 percent, and 90 percent of a specific group (such as the elderly or persons with a heart condition) would experience a particular symptom. This was done for three levels (discomfort, incapacity, disability) for each of fifteen groups of individuals (normal and sensitive populations) for carbon monoxide, oxidants, and nitrogen dioxide. Where there was a large spread in the estimates, the experts were given a second chance to make an estimate after being told what the others had said, which reduced the variability somewhat. Sigmoid dose-response curves then were fitted to the estimates.

The dose-response function for persons suffering a heart condition, angina pectoris, is shown in figure 8-1. For comparison, the dose-response function developed by Nelson, Knelson, and Hasselblad (1976) for the same effect also is shown. Gillette (1977) has utilized the Leung, Goldstein, and Dalkey (1978) dose-response functions and his own estimated values of discomfort, incapacity, and disability to estimate a damage function for ozone health effects.

Valuation of Health Damages

Air-pollution health effects may be valued either in terms of a unit change in health—for example, a day of illness or a human life—or in terms of changes in the risk of becoming ill or of dying. In general, the most-appropriate measure for illness is the value per day of illness, and the most-appropriate measure for life is the value of a change in the statistical risk of dying.[4]

The valuation of effects on health should be based on the preferences of individuals, as revealed by their willingness to pay to avoid the effects or by the amount they must be paid to compensate them fully for the effects. The amount they are willing to pay is analogous to the compensating-variation measure of consumer's surplus, and the required compensation is analogous to an equivalent-variation measure. As discussed in chapter 6, if the changes in the resulting health effects are small, these two measures will yield similar results.

Estimates of the required compensation for, or willingness to pay to

Estimating Health Benefits

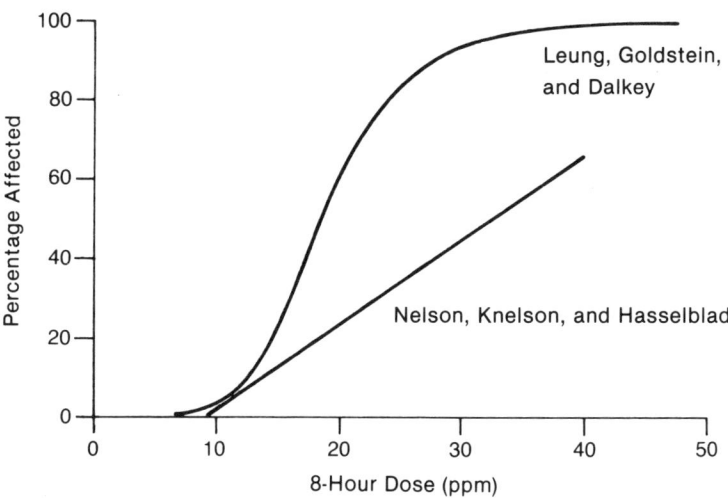

Source: Steve Leung, Elliot Goldstein, and Norman Dalkey, *Human Health Damages from Mobile Source Air Pollution: A Delphi Study,* vol. 1, EPA 600/5-78-016a (U.S. Environmental Protection Agency, 1978); Calculated from data in William C. Nelson, John H. Knelson, and Victor Hasselblad, Air pollution health effects estimation model, in *Proceedings of the Conference on Environmental Modeling and Simulation,* EPA 600/9-76-016, pp. 191-195 (U.S. Environmental Protection Agency, 1976), using data in J.E. Peterson and R.D. Stewart, Predicting the carboxyhemoglobin levels resulting from carbon monoxide exposures, *Journal of Applied Physiology* 39(1975):633-638.

Figure 8-1. Subjective Dose-Response Curves for Excess Angina Attacks from Exposure to Carbon Monoxide

avoid, an extra day of illness or an increase in the risk of dying can be obtained directly from public surveys or indirectly from actual prices of other goods and services. The validity of survey data is limited because people are asked to speculate on the hypothetical payment they would make for something they do not normally purchase. Other aspects of the interview process raise additional opportunities for bias to affect the results.

Estimates of the amount individuals are willing to pay to avoid several common symptoms of air-pollution health effects were obtained from a survey in the Tampa, Florida metropolitan area by Loehman et al. (1979). For example, they report that the median individual would be willing to pay $3.80 to avoid one day of minor eye, ear, nose, and throat irritation. These amounts were interpreted as reflecting only inconvenience, pain, and suffering. The total costs of an illness also will include direct medical costs and lost income. Ruby (1978) has developed estimates of these costs for an average case of several illnesses that frequently are associated with air pollution. These estimates are presented in table 8-3.

Surveys also have been conducted to learn the value of changes in the risk of dying (Acton 1973; Jones-Lee 1976). However, these studies involved very small samples; and the risks people were asked to evaluate are difficult to interpret in terms of more-general environmental health risks.

Compensating Wage Premiums

Market data on the wage rates for jobs involving different degrees of risk to life or health have been used to infer the wage premiums required to compensate individuals for accepting additional risks. In order to isolate the compensating wage premiums from other influences on wages, multiple-regression techniques have been used to estimate equations for total wages paid that include variables for personal characteristics (such as age, race, education) and job characteristics (such as hours worked, occupation, unionization) as well as the degree of risk. The economic models underlying estimates of the compensating wage premium are discussed by Linnerooth (1979), Freeman (1979b), and Thaler and Rosen (1976). Reviews of the empirical results have been published by Smith (1979) and Bailey (1980).

The compensating wage premiums inferred from the market data are not necessarily equal to the amount an average individual would require to be compensated fully for accepting additional risks. Individuals differ in their attitudes toward risk, with those who are less risk averse accepting the more-hazardous jobs. The actual premiums paid will depend on the supply of workers at the wage premium offered and the cost of providing greater job safety.

Thaler and Rosen (1976) estimated the compensating wage premium for a sample of workers in hazardous occupations. The workers in their sample included a higher-than-average proportion of minority and otherwise economically disadvantaged workers. No variable was included for injuries, so the estimated fatality coefficient will be biased to the extent that fatalities and injuries are correlated. Their estimates of the marginal wage premium for a change in the risk of fatalities of 1.0×10^4 range from \$34 to \$51 per year.[5] The average risk of fatalities in their sample was 11.0×10^{-4}, compared to a risk of 0.9×10^{-4} faced by the average U.S. worker.

Brown (1980) used the same source of data on occupational fatality rates as did Thaler and Rosen, but used a different sample of workers. The average fatality risk in his sample was 2.2×10^{-4}. Brown's estimates of the marginal wage premium are approximately three times larger than those of Thaler and Rosen. Brown attributed the difference to the greater number of variables included in his model rather than to the inclusion of less-risky occupations in his sample.

Viscusi (1978) and Olson (1981) estimated the marginal wage premium using data for risks by industry rather than by occupation. The use of indus-

Table 8-3
Estimated Expenditures on Illness

Illness	Period	Medical Costs	Lost Income
Aggravation of heart and lung symptoms	Daily	$ 22	$ 12
Aggravation of asthma	Daily	29	21[a]
Juvenile lower-respiratory disease	Case (5-15 days)	120	85[a]
Adult chronic respiratory disease	Annual	1,400	1,400

Source: Michael G. Ruby, An application of benefit-cost analysis: The ASARCO-Tacoma copper smelter, Paper presented to the Annual Meeting of the PNWIS—Air Pollution Control Association, 1978 (Paper PNWIS 78-12).
Note: In 1978 U.S. dollars.
[a]Includes lost income of adult when direct care of juvenile is required.

trial data allowed them to include variables for injuries as well as fatalities in their regressions. The average risk in their samples is approximately equal to the average risk for all U.S. workers.

Viscusi's estimates of the marginal wage premium range from $230 to $280. He attributes his higher estimates to the inclusion in his sample of the more-risk-averse workers in the less-risky industries. Viscusi also obtained an estimate of the marginal risk premium for injuries of about $1. However, this estimate is difficult to interpret because of the wide range of types of injury represented by the injury variable.

Olson included both risk and risk-squared in his equations to allow the estimated marginal wage premium to vary with the degree of risk.[6] His results indicate that the total wage premium increases more slowly than total risk, implying that the marginal wage premium decreases as risk increases. This is consistent with the pattern of results from the other studies, in which lower marginal wage premiums are estimated with data for workers exposed to higher levels of risk. The expressions Olson obtains for the marginal wage premium are in the form $a - b\Pi$, where Π is the probability of a fatality multiplied by 10^4. His estimates range from $(548 - 75\Pi)$ to $(560 - 71\Pi)$. Olson was able to reproduce Viscusi's estimates with his data by deleting the risk-squared term in his regressions.[7]

The injury variable included in Olson's equation measured the severity of accidents by the average number of work days lost. His results indicate a marginal premium of $1 for an accident and an additional premium of $0.50 for each day lost from work. Thus the marginal wage premium for a two week absence from work would be $6 and the premium for an accident resulting in a full year away from work is about one-fourth the fatality wage premium. This ratio is consistent with the subjective evaluation of disease states reported by Kaplan, Bush, and Berry (1976).

Adjustments to Wage Premiums

Estimates of required compensation for risk from studies of compensating wage premiums reflect the value of risks voluntarily assumed by taking a particular job. Health risks from air pollution, however, are involuntary. Slovic, Lichtenstein, and Fischhoff (1979) provide evidence that involuntary risks may be more costly. They asked a panel of adults to assign relative social values to deaths from a number of causes. Deaths due to one's own personal choice or negligence (for example, while mountain climbing) were assigned values near 1.0. Deaths from common, voluntarily accepted risks (such as riding in automobiles) were rated about 1.25. Deaths from involuntary and insidious risks (such as exposure to pesticides) received a rating from approximately 1.5 to 2.0. This suggests that data from studies of occupational risk may understate the required compensation.

Also, if the compensating wage premium did not fully reflect the distress and inconvenience caused the relatives and friends of the employee the wage premium would be an underestimate of the true costs.[8] Thaler and Rosen (1976) found that a higher wage premium is paid to married men, which suggests that workers are considering the effects on others. It is not possible to know if the additional amount demanded by a worker understates or overstates his actual worth to his family and friends.

The compensating wage premium may overstate the value if the social value of the reduction in risk is related to the number of years of life saved, as suggested by Zeckhauser and Shepard (1976) and Schwing (1979). The estimated wage premiums refer only to individuals of working age, and may overstate the value for the average (older) individual vulnerable to air pollutants.

These considerations do not present a clear picture of either an understatement or an overstatement of the required compensation by the marginal wage premium estimates. It is suggested that no adjustment should be made to the reported values.

Human Capital

In order to be consistent with the theoretical basis for benefit-cost analyses discussed in chapter 2, values of life and health should be based on the preferences of those affected. However, procedures that ignore these preferences frequently have been used in empirical studies.

The most common value used for loss of life has been the present value of the income the individual would have received had he lived. The rationale for this approach is that individuals are paid according to the value of their output and that, therefore, the present value of their future income is equal to the present value of the output they would have produced. In effect, an individual is valued as if he were a machine, whose value to society lies only in what it is able to produce. Despite its dubious rationale, this approach has been quite common, perhaps as a result of the ready availability of the data required to implement it.

The best-known computations of foregone earnings are those of Cooper and Rice (1976). They considered both lifetime-earnings patterns and the expectation of survival from initial age. An implicit income for homemakers was estimated to account for nonmarket activities. They report discounted present values of expected foregone earnings specific to sex, race, and age with discount rates of 4 percent and 6 percent. By comparing the expected age at death for an average population to that for a sample of persons affected by the proposed policy or project, it is possible to estimate the total market and nonmarket future labor value lost due to premature death. For example, if Cooper and Rice's lifetime-earnings data are used with the age- and sex-specific coefficients for excess deaths associated with sulfate air pollution calculated by Mendelsohn and Orcutt (1977), a present value of $68,000 is implied for the "average" individual.

Summary

Damage functions for health effects require both a dose-response function for the effects on health of changes in pollution concentrations and value factors stating the amount the effects are worth to society. Although both these items are poorly understood at present, data have begun to accumulate and rough estimates of damage functions can be made for some effects from most of the common pollutants.

Laboratory and microepidemiologic studies of pollutant effects seldom provide dose-response functions that are useful for estimating the economic value of controlling air pollution. Data from macroepidemiologic studies are more easily translated into useful damage functions, but are by their nature much less precise. Because many factors influence the health of individuals, it is necessary to use statistical methods to isolate the effects of air pollution. The wide possible choice of variables and data sources is a primary cause of disagreement among analysts. Several otherwise good studies are badly flawed by poor data on air-pollution exposures. Despite these problems, several studies of the variation in mortality rates across cities have reported a significant relationship with air-pollution concentrations.

Recognizing the difficulty in establishing objective dose-response functions, a few analysts have used data from the experiments and the subjective opinions of health-effects researchers to develop dose-response functions for several pollutants and their effects.

The values assigned to health effects should be based on individual preferences. The common use of the present value of foregone earnings does not provide a conceptually adequate estimate of the values. A promising source of information on the amounts required by individuals to compensate them for accepting additional risk are the statistical studies of compensating wage premiums.

The results of the compensating-wage-premium studies cited here suggest an average social value for a change in risk of 1×10^{-4} between \$250 and \$500 per capita. If we accept the mortality-exposure relations suggested by the macroepidemiologic studies discussed earlier in this chapter, these risk values give a per capita value of the reduction of the annual average concentration of sulfur dioxide or total suspended particulate by one $\mu g/m^3$ of approximately \$10 to \$25. The per capita value of a reduction in the annual average concentration of sulfate by one $\mu g/m^3$ is higher, approximately \$75 to \$150, but even more uncertain.

Notes

1. However, including socio-economic variables in the estimating equations will partially control for such adjustments. Therefore, an estimate closer to the distance $Q_2 - Q_0$ in figure 6-5 may be obtained.

2. Lee, Caldwell, and Morgan (1972) and others have failed to find a single, consistent conversion between the COH measure and direct mass measurements such as TSP. However, their data suggest a conversion of 1 COH to 150 $\mu g/m^3$ TSP as a useful approximation within the range of the data in these studies.

3. Because sulfate is a major constituent of the fine-particulate fraction and sulfate readings in the United States are obtained by chemical analysis of the TSP filters, sulfate measures may be highly correlated to the fine-particulate fraction of TSP. Other chemical species, such as aluminum and iron, are closely associated with the coarse-particle fraction. Use of several such elements as variables may result in high multicollinearity and low statistical significance for the estimated coefficients.

4. Changes in risk are measured by changes in the probability of the occurence of the event. An increase in the risk of fatalities of 1.0×10^{-4} corresponds to an increase in the expected number of fatalities of 1 per 10,000 individuals in the relevant group (for example, employees in a particular industry).

5. All estimates of marginal wage premiums reported in this section are expressed in terms of 1978 U.S. dollars and refer to changes in risk of 1.0×10^{-4}.

6. Thaler and Rosen (1976) also included a risk-squared variable in preliminary regressions but found it was not significant for their sample.

7. For the extremely hazardous occupations studied by Thaler and Rosen (1976), Olson's (1981) estimated relationship between the marginal wage premium and average level of risk would imply negative premiums.

8. Needleman (1976) used data on the behavior of kidney donors in the United States to estimate the costs to family and friends. These data provide direct estimates only for the immediate family, but he extrapolates to a value for close friends. The value for all others is assumed to be zero. If Needleman's estimates are summed using the 1980 family and population characteristics of the United States, then the cost to others is estimated to be 44 percent of the value an individual places on his or her own life.

9 Estimating Vegetation and Ecosystems Benefits

Numerous studies have demonstrated that air pollution can adversely affect the appearance and growth of plants and animals. In an economic analysis it is necessary to focus on the specific effects that result in a loss of social value. For agricultural crops and livestock, the relevant effects are the reductions in the quantity and quality of marketable products. For ornamental vegetation, losses take the form of reductions in aesthetic value caused by visible injury to the plant. The relevant effects on ecosystems are less clear, because not enough is known about all the ways in which air pollution affects natural processes. A review of the economic assessment of air-pollution damage to vegetation has been published by Benedict and Jaksch (1979).

Most studies of air-pollution effects on vegetation focus on the visible or biological changes in the plant following an exposure to air pollution rather than on the economically important consequences of the exposure. Careful use of terminology will keep this distinction clear. Several analysts have suggested using the term *injury* to describe changes of little or uncertain relation to the economically important part of the plant and reserving the term *damage* for effects that result in economic loss.

This chapter reviews the available information that may be helpful in estimating air-pollution damages to vegetation and ecosystems. Very few dose-response functions for reduction in marketable yield have been published. Therefore, some of the dose-response functions that have been developed for visible injury and the damage factors that have been used by several analysts are described. The chapter concludes with a discussion of the assignment of economic values to observed vegetation and ecosystem damages.

Vegetation Dose-Response Functions

The quantity and quality of agricultural products depend on numerous conditions, including temperature, light intensity, humidity and rainfall, and soil nutrients. The effects of air pollution on plants also will depend on weather and soil factors, on the specific variety of plant, and even on the past history of the individual plant. Many plant species have been assigned to "sensitive," "intermediate," or "resistant" classes with respect to pollutants. A summary of such assignments for most pollutants and many North American plants can be found in Lacasse and Treshow (1976). Detailed lists are available for specific plant groups (for example, Davis and Wilhour 1976) and other geographic areas (for example, O'Connor, Parbery, and Strauss 1974, 1975). Relatively complete assignment lists can be found in some critical reviews for individual pollutants (for example, Heck 1977).

Because many of the variables can be controlled and dozens of almost identical plants can be grown for use in a laboratory experiment, there have been numerous laboratory studies of the effects of air pollutants on plants under varying conditions. However, most of the responses reported in these studies are difficult to relate to economic losses. Surprisingly, marketable-yield reduction is rarely measured. In the few studies that measure yield reduction, this often is done only at one or two concentrations of the pollutant.

For some crops and pollutants (such as fluoride), the buildup of the pollutant concentration in the plant, usually measured as parts per million (ppm) by weight or the numerically equal micrograms of pollutant per gram of plant ($\mu g/g$), is more frequently reported as the pollutant variable than is the concentration of the pollutant in the air. To construct a dose-response function in these circumstances, it is necessary to locate additional studies relating airborne-pollutant concentration and duration of exposure to the measured pollutant burden in the plants. However, this introduces significant additional uncertainties.

Where the economic analysis can be cast into a choice between no pollution and a specific level of pollution, it may be possible to locate a few studies that have measured yield loss at approximately the pollution level in question. Reviews of air-pollution damage to plants (for example, Ormrod 1978; Guderian 1977) or critical reviews of a specific pollutant are helpful in locating such studies. For other changes in the pollution level, a dose-response function is necessary. However, only a few experiments with enough information to define directly a dose-response function for marketable yield have been published. They are described later on. A few other dose-response functions have been constructed by assembling and analyzing previous studies, and these also are discussed later in the chapter.

Yield Dose-Response Functions

Two of the dose-response functions for marketable-yield loss were developed in the Los Angeles, California basin for the effects of ozone. The dose was defined in each study as the sum over the growing season of the hourly average concentrations (expressed in parts per hundred million, pphm) greater than 10 pphm ozone. Oshima et al. (1976) used data from nine field plots to obtain a function for the percentage reduction in harvestable yield of alfalfa (variety Moapa 69). The estimated dose-response function is

$$I = -1.1 \times 10^{-4} + 9.3 \times 10^{-3} D \qquad (9.1)$$

where I is the percentage reduction in yield and D is the total dose, as defined previously. The coefficients are statistically significant at the 99-percent confidence level. Oshima et al. (1976) also present several equations for yield that include temperature and relative-humidity variables. The coefficient of the dose variable in equation 9.1 reflects the inclusion of these environmental variables. Although their dose-response function is specific to a particular variety of alfalfa, it approximates the yield losses observed in other experiments with different varieties of alfalfa.

Oshima et al. (1977) report a dose-response function for the harvestable yield of fresh market tomatoes (variety 6718 VF). Their estimated equation is

$$I = 0.023 \, D \qquad (9.2)$$

where I is the percentage reduction in yield and D is the dose, as described previously. The coefficient is statistically significant at the 99-percent confidence level. Temperature and relative humidity were found to be much less important in determining the yield loss and were not included in estimating the dose-response function. Oshima et al. observed that this simple relation tended to understate the impact on the farmer, since the ozone exposures reduced the size and value of the fruit as well as the total yield, and delayed much of the harvest into the later part of the harvest season when prices were lower. Other varieties of tomato have shown similar response patterns in other tests.

A group of studies of ozone damage conducted in North Carolina may suggest harvestable-yield dose-response functions appropriate to the eastern United States for several crops. For these studies, the dose was defined as the average ozone concentration during the growing season between 9:30 a.m. and 4:30 p.m. each day. No dose-response functions were constructed from the data, but the three concentrations used cover the entire range of expected exposures. The results are summarized in figure 9-1. At least ten varieties of each crop were included in a preliminary screening; the final tests were made with several varieties of varying degrees of sensitivity.

Visible Injury

Most studies of vegetation injury by air pollutants measure the effect by the extent of injuries to leaves or changes in photosynthesis rates. Heck (1977)

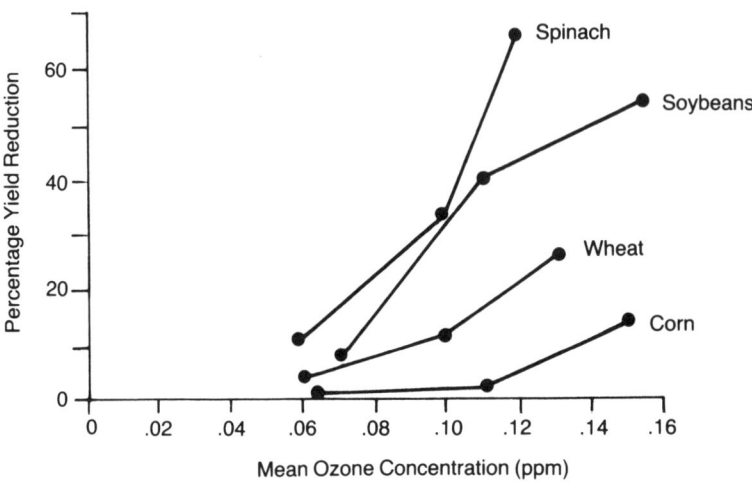

Source: Spinach: Allen S. Heagle, R.B. Philbeck, and Michael B. Leckworth, Injury and yield responses of spinach cultivars to chronic doses of ozone in open-top field chambers, *Journal of Environmental Quality* 8(1979):368-373; Soybeans: Allen S. Heagle, A.J Riordan, and Walter W. Heck, Field methods to assess the impact of air pollutants on crop yields, paper presented to the Annual Meeting of the Air Pollution Control Association, 1979 (Paper 79-46.6); Wheat: Allen S. Heagle, Suzanne Spencer, and Michael B. Leckworth, Yield response of winter wheat to chronic doses of ozone, *Canadian Journal of Botany* 57(1979):1999-2005; Corn: Allen S. Heagle, R.B. Philbeck, and W.M. Knott, Thresholds for injury, growth, and yield loss caused by ozone on field corn hybrids, *Phytopathology* 69(1979):21-26.

Figure 9-1. Reductions in Yield for Four Crops

and Larsen and Heck (1976) used data gathered by many researchers to estimate dose-response functions for leaf injury caused by ozone. Heck (1977) estimated equations of the form

$$C = A_0 + A_1 I + A_2/t,$$

where C is the concentration of ozone (in pphm), I is the percentage of leaf area reported as destroyed, and t is time (in hours). The estimated values of the coefficients A_0, A_1, and A_2 are listed in table 9-1. The equation assumes that the concentration is constant over the time period. It is not assumed to be valid for periods longer than twelve hours. Heck also reports coefficients for several additional groups of plants.

Larsen and Heck (1976) analyzed data for ozone leaf injury to fifteen plants and sulfur-dioxide leaf injury to four trees using equations of the form

$$C = mt^p s^z,$$

where C is the concentration of pollutant; t is time; m, p, and s are the esti-

Table 9-1
Estimated Coefficients for Heck Ozone Dose-Response Functions

Plants	A_0	A_1	A_2
Sensitive			
All plants	−1.52	0.40	21.3
Grasses	−5.65	0.48	29.1
Tobacco	2.45	0.34	13.7
Intermediate			
All plants	2.44	0.65	29.0
Grasses	1.87	0.59	29.2
Clover	−0.99	0.71	26.8
Tobacco	6.31	0.87	15.2
Resistant			
All plants	16.89	0.95	27.8
Grasses	19.06	1.17	26.3

Source: [Walter W. Heck], Plants and microorganisms, in *Ozone and Other Photochemical Oxidants,* pp 437–585 (Washington, D.C.: National Academy of Sciences, 1977).

mated coefficients; and z is related to the measured percentage of leaf injury. The coefficients are reported in their paper. These equations also assume that the concentration is constant over the time period and that the duration is less than twelve hours.

A benefit-cost analysis generally will require an estimate of the effects as a function of concentration. However, the equations of Heck (1977) and Larsen and Heck (1976) were estimated with the pollutant concentration expressed as a function of the observed injury. The inverse of the estimated coefficient of the injury variable in their equations will not be equal to the coefficient of the concentration variable in a regression expressing injury as a function of concentration.

It would be tempting to use these acute leaf-injury dose-response functions as dose-response functions for yield reduction, particularly for plants such as tobacco or spinach, where the leaf is the economic portion of the plant, or for ornamentals, where visible injury reduces the aesthetic value. However, the effects on yield or value may be substantially different. One problem is that these leaf-injury functions are specifically developed to estimate damages from acute exposures. It would be possible to sum the daily exposures across a growing season using an equation similar to 8.1, but such a sum may not correctly represent the effects over a growing season. The time at which the injury occurs can be an important factor. Leaf injury to a young plant or to a young leaf may affect the entire growth of the plant, reducing the yield even more than the sum of the specific leaf-area injuries. Also, for some leaf crops, damage to only 40 percent of the leaf area may make the crop 100 percent unmarketable.

Several studies have attempted to relate the amount of leaf injury to the reduction in yield. Early work on the exposure of alfalfa to sulfur dioxide (Hill and Thomas 1933) found that, for a single exposure of one to two hours, the eventual yield reduction was approximately one-third the percentage of leaf area injured. When two exposures occurred during the growing season, the yield reduction was about one-half the total percentage of leaf area destroyed. When the alfalfa plants were exposed three times—at early, middle, and late stages of growth—the yield reduction was about three-fourths the total percentage of leaf area destroyed. Similar results were obtained by cutting off portions of the leaf. Another study (Davis 1972) found the percentage yield reduction in soybeans was about two-thirds the percentage of leaf injury for multiple exposures.

Some studies have suggested that a reduction in yield can occur even when there is no visible injury. For example, long-term exposures to sulfur dioxide have been reported to reduce significantly the yield of pasture grasses (Crittenden and Read 1978) and alfalfa (Tingey and Reinert 1975) without any visible injury present. However, growth and yield loss generally have been assumed to be related to visible injury.

The leaf injury and yield losses predicted by experiments conducted in the laboratory may be different from those observed in field studies. In the laboratory, plants are usually exposed to a constant concentration of the pollutant and ideal conditions of soil, water, temperature, light, and so forth. Some experiments have suggested that plants are most susceptible to pollution injury when the ideal amounts of light, temperature, and nutrients are available. Plants growing in the field are subject to varying concentrations of pollutants and to fluctuations in other variables. Greater movement of the air in the field will increase the actual contact between the plant and the pollutant. Naturally growing plants also may be growing in the extremes of their biological range and may thus be very sensitive to any additional stress such as air pollution. Because of these contradictory tendencies, it is not clear whether results from laboratory studies will over- or underestimate the effects that would be observed under more natural conditions.

Loss Factors

As part of an attempt to estimate national vegetation damages from air pollution, Benedict, Miller, and Smith (1973) developed several subjective dose-response functions for a wide variety of plants. They divided most of the important agricultural crops and ornamentals into five sensitivity classes each for ozone, sulfur-dioxide, and fluoride exposure, based on the air-pollution susceptibility of the plant and the economic importance of the part of the plant affected by the pollutant. A table of loss factors was established for various concentrations of pollution for each class from various measurements of plant damage from these pollutants. The loss factors were estimates of the fractional reduction in yield from the economically valuable portion of the plant due to pollution.

Because reliable ambient-pollution data were available for very few areas in 1970, Benedict, Miller, and Smith estimated the amount of potentially damaging ozone pollution from the emissions of hydrocarbons and nitrogen dioxide in each county and from Holzworth's (1972) estimates of the average number of local two-day atmospheric stagnations during the growing season. Although it is conceptually satisfying, this estimate of pollution exposure is only weakly correlated to actual ozone concentrations measured a few years later ($r = 0.2$).[1] Because sulfur-dioxide and fluoride concentrations tend to decrease fairly quickly with distance from a source (unlike ozone concentrations, which can be reasonably uniform across large areas), their estimates of sulfur-dioxide- and fluoride-pollution potential were based on the number of large sources of these pollutants in each county rather than on total emissions. The correlation between the estimated sulfur-dioxide-pollution potential and subsequently measured sulfur-dioxide annual average and peak concentrations is moderate ($r = 0.4$).[2]

Liu and Yu (1976) used the county-by-county vegetation-damage estimates made by Benedict, Miller, and Smith in estimating damage functions for ozone and sulfur dioxide. They used both the Benedict, Miller, and Smith pollution-potential estimates and 1972 annual average sulfur-dioxide concentrations as their pollution variables. They found a significant correlation between the pollution potential estimates and the crop loss estimates, but since the latter were generated from the former, this result is not very meaningful. They did not find a statistically significant correlation between measured sulfur dioxide concentrations and the crop loss estimates. Considering the weak correlation of the Benedict, Miller, and Smith oxidant pollution potential estimates with oxidant exposures, mentioned above, and the statistical insignificance of the sulfur dioxide measurements, the Liu and Yu dose-response functions should not be relied on for crop damage estimates.

Field Surveys

Actual surveys of crop damage have been attempted in several states (for example, Millecan 1976). To conduct such a survey, field workers (often the county agricultural-extension agent) are first trained to recognize the visible symptoms of air-pollution damages. They then make a farm-to-farm survey, estimating the percentages of each crop showing various categories of damage. These totals for visible damage are subjectively translated into a marketable-yield loss. A detailed description of the procedures for conducting such a survey has been given by Skelly, Krupa, and Chevone (1979). Because the surveys are necessarily selective and the yield conversions poorly understood, the values obtained are only estimates. Nevertheless, they often represent the best available information on crop losses. In California the estimates have been fairly stable from year to year. In the eastern United States, where weather is more variable, the estimates change more from year to year.

Animal and Ecosystems Dose-Response Functions

Although adverse effects of air pollutants are frequently demonstrated on laboratory animals, there are very few reported incidents of pollution damage to livestock or wildlife except from fluorides. Consistent adverse effects on milk production and weight gain in cattle are observed at high fluoride exposures, but dose-response functions cannot be generalized from the information available. Adverse effects on the teeth and bones of cattle are observed at lower fluoride exposures, and economic damages result from increased veterinary expenses or reduced marketability of the animals. Published reports concentrate on describing the injuries, rather than estimating the economic loss. Thus damage claims often provide the only source of information, and these often are considered confidential if an out-of-court settlement has been achieved. A farm-by-farm review usually is necessary to establish the magnitude of the economic damage to livestock.

Air-pollution effects on the smaller parts of the ecosystem are also difficult to evaluate. Some information is available on the body weight of absorbed pesticides or metals that will adversely affect bees and other insects, although it is very difficult to relate such data to atmospheric exposures. Limited information also is available on the concentrations of pollutants that will increase plant sensitivity to insect damage (which may alter the succession of plant communities), the concentrations of sulfur dioxide that will kill lichens (which affects long-term soil creation), and the buildup of acidity from acid rain (which can result in the elimination of entire aquatic plant and fish populations from lakes and streams). With the exception of acid-rain effects, which have been studied extensively in Scandinavia, very little is known about the long-term effects of such ecosystem damage. Some observations have been made of changing species composition of forest and pastures and some research is being conducted on effects on climate and the upper atmosphere. Because the effects are so poorly understood and appear to require long exposure periods before they become observable, there are no dose-response functions available for ecosystem injury from air pollution.

Valuation of Vegetation, Animal, and Ecosystem Damages

The economic damage to vegetation from air pollution will depend on the change in quantities and prices of marketed commodities and the loss of consumer enjoyment in ornamental plantings. Local changes in air-pollution concentrations will have little effect on the prices of crops grown by geographically diverse producers who face national markets (such as wheat or corn), although the prices of other crops (such as fresh produce) may be very sensitive to changes in local supplies. Using the current (average) price of the commodity to estimate the loss in value generally will result in an overestimate of the damage from increased pollution or an underestimate of the benefits from reduced pollution. The amount of the error will depend on the demand and supply characteristics of the particular crop, as discussed in chapter 6.[3] In some cases the error may be substantial. In such cases a more-detailed analysis of price movements and of changes in planting patterns as farmers respond to the new prices will be necessary.[4]

If air pollution affects the quality or timing of harvest of the crop as well as the quantity, this may also reduce the price. Such reductions in price, which reflect lower values of the commodity to consumers, should be included in calculation of the losses from air pollution.

The use of current crops to estimate the benefit from a reduction in pollution may result in an underestimate since farmers already may have made partial adjustment to the pollution by substituting crops that are more resistant but less profitable than the crops they could grow in the absence of pollution. In the Los Angeles basin at least twenty crops (including tomatoes, spinach, and flowers for seed) can no longer be grown profitably (Millecan 1976). This results in a greater revenue loss to the farmer and an inefficient allocation of agricultural land.

Some analysts have used various multipliers to increase the revenue loss to the farmer to reflect the eventual price paid by the consumer. This is not a correct procedure. The demand for agricultural products is derived from the demand for finished goods or fresh produce at a convenient location. Price markups at intervening stages of processing reflect returns to those services (for example, milling of flour or baking of bread) that occur regardless of the source of the original agricultural product. If the change in quantity available is large enough to affect the structure of the processing industries or to threaten the continued existence of farms, then significant readjustments may occur. However, calculation of losses will require detailed analysis of the changes in the industry as readjustments are made, rather than the use of simple multipliers.

The valuation of damages to ornamental plants requires a different approach than is used for agricultural crops. The principal value of ornamental plants is in their daily enjoyment by owners and observers, which may be reduced by visible injury from pollution. Thus ornamental plants are a capital asset providing a flow of aesthetic services. If the size of the capital stock is not changing, the annual maintenance and replacement expenditures represent an approximate estimate of the annual value of these services. From a mapping of land in residential uses, public buildings, parks, and highways, Benedict, Miller, and Smith (1973) estimated that the average annual expenditures on ornamental plantings was more than $93 per household (in 1978 U.S. dollars) in the approximately 500 more-polluted counties in the United States. The loss in value will be a function of both the amount of foliar injury and the reaction of the public to the injury. The general issue of valuing aesthetic damages is discussed in chapter 11.

The observed injuries to livestock relate mostly to high concentrations of pollutant in the forage. In some instances farmers will rotate their animals between contaminated and clean fields or supplement their own feed with uncontaminated feed from elsewhere in order to keep the accumulated pollutants in the animal below critical levels. Thus the economic damage to livestock is generally found in the increased costs of avoiding the effects of the air pollutants, rather than in direct losses of animals or milk production. The reduced value of the land that might otherwise be used to pasture livestock also could be used in making such estimates, but care must be taken to avoid double-counting.

Valuation of damages to ecosystems is especially difficult. Westman (1977) has made a few illustrative calculations of the economic damages of some long-term effects to ecosystems that may suggest some approaches to the problem, although adequate information simply is not available to generate estimates for any but the most-severe situations. Several approaches to the calculation of costs associated with long-term changes in climate are reported in Ferrar (1976), but the association between changes in pollution concentrations and changes in climate is too poorly understood for these procedures to be very useful in a benefit-cost analysis.

Summary

Derivation of a damage function for agricultural commodities requires a dose-response function relating changes in marketable yields to changes in pollution concentrations. Unfortunately, most existing dose-response functions measure effects such as leaf injury rather than changes in yield. Attempts have been made to relate leaf injury and yield loss, but the results should be viewed only as rough approximations.

Valuation of changes in yields for agricultural commodities can be based on data on prices and quantities sold. However, valuation of damages to ornamental plants requires estimates of the value of lost aesthetic ser-

Vegetation and Ecosystems Benefits

vices. Little information is available on which to base damage functions for animals and ecosystems.

Notes

1. The measure of ozone exposure used to estimate the correlation was the number of hours in which ozone concentrations exceeded 0.08 ppm, as reported in U.S. Environmental Protection Agency (1976).

2. Annual average and 90th percentile of sulfur-dioxide concentrations are reported for these counties in U.S. Environmental Protection Agency (1972).

3. Demand elasticities for various agricultural crops in the United States are reported in George and King (1971). Caspari, MacLaren, and Hobhouse (1980) report demand and supply elasticities for European countries. Supply elasticities for specific crops in specific geographic areas have been published in reports of state agricultural agencies or university agricultural-economics departments.

4. Leung et al. (1978) and Adams, Thanavibulchai, and Crocker (1979) provide discussions of the use of econometric models to make estimates of the adjustments of farmers and consumers to substantial changes in quantities and prices.

10 Estimating Materials Benefits

Many materials with economic value can be damaged by air pollution. The most common form of damage is corrosion of exterior surfaces by acidic conditions, primarily originating from sulfur-oxide emissions. Rubber and other elastomers crack and lose resiliency when exposed to ozone. Nitrogen dioxide and ozone accelerate the fading of fabric dyes, and sulfur-dioxide exposure can reduce the strength of textile fibers. Aesthetic losses from soiling by particulate matter will be discussed in the next chapter.

The principal economic loss from materials damage is a shortened service life. Either maintenance activities or replacement must occur sooner than would be expected in the absence of pollution. The economic value of the loss will be the difference between the present value of the earlier replacement and that of the later replacement. If the damaged material is a component of a larger item that may fail or require replacement as a unit, then the loss must reflect the total cost of repair or replacement.

From a review of six broad studies of air-pollution damages to materials, Yocom and Grappone (1976) developed an estimate of the relative economic importance of the pollutants and the materials they affect. Their results are shown in table 10-1. Each entry is the percentage of the total economic loss from air-pollution damage to materials that can be assigned to that specific pollutant-material interaction. The entry for particulate damage to paints includes aesthetic losses caused by soiling.

A comprehensive review of damage to materials by Salmon (1970) identified (in order of importance) metals, cotton, paint and coatings, building stone, paper, and leather as the primary materials damaged by sulfur dioxide. Paint and coatings, building stone, and glass were found to be the materials most often damaged by particulate matter (primarily by soiling). His findings were similar to those in table 10-1 for other pollutants and materials.

This chapter presents dose-response functions for changes in the service life of steel, zinc, painted surfaces, and cotton textiles. Data are reported for the distribution and value of these materials in the economy. A method for expressing the damages as an annual cost is described.

Table 10-1
Relative Economic Importance of Pollutants in Materials Damage

Material	Sulfur Dioxide (%)	Particulate Matter (%)	Ozone (%)	Nitrogen Dioxide (%)	Total (%)
Paints	9	23			32
Metals	18				18
Rubber			22		22
Textiles			4	6	10
Other	14	4			18
Total	41	27	26	6	100

Source: Adapted from John E. Yocom and Nicola Grappone, *Effects of Power Plant Emissions on Materials,* EPRI EC-139 (Palo Alto, Calif.: Electric Power Research Institute, 1976). 1976).

Metals Dose-Response Functions

Metal surfaces exposed to the atmosphere are subject to natural corrosion whenever there is enough water vapor to wet the surface. The presence of pollution in the air, particularly of sulfur dioxide, will both reduce the critical level of relative humidity where wetting occurs and increase the rate of corrosion.

Iron and steel, aluminum, copper, lead, and zinc are the most widely distributed metals in the economy. Together they account for more than 85 percent of the economic value of metals in use. Iron and steel represent about half of the economic value and are the most susceptible of these metals to corrosion damage. Aluminum pits and corrodes to a limited extent and sulfur-dioxide pollution increases the corrosion, but the damage appears to be slight and of no economic significance. Copper also is subject to surface corrosion, which produces a green patina finish that is aesthetically desirable in exterior applications. The formation of this surface coating in turn slows down the corrosion process substantially. Lead is almost totally resistant to corrosion, even in polluted air. Zinc is widely used as a sacrificial corrosion-protection coating, with about half of the total zinc production used in galvanized-steel production. The erosion of the zinc protective-surface layer is considerably accelerated by sulfur-oxide pollution.

Protection from corrosion problems can be obtained by substituting a metal such as aluminum or by providing a protective coating of plastic, paint, or zinc galvanizing. The cost of galvanizing is about fifty times less than that of painting, so a considerable fraction of exposed steel surfaces

Estimating Materials Benefits

is galvanized. Many galvanized surfaces are painted for additional protection or for aesthetic reasons. Since both paint and zinc can be eroded by sulfur-oxide pollution, the protection is only temporary.

Steel

Air-pollution damage to steel can be a more-severe problem than the simple erosion of a surface. It can contribute to a reduction in strength by accelerating stress-corrosion cracking. For example, the failure of the Highway U.S. 35 Bridge over the Ohio River in 1967, which killed forty-six persons and injured nine others, was traced to the failure of a single structural element by stress-corrosion cracking fatigue caused by a sulfur-containing air pollutant, possibly hydrogen sulfide (Gerhard and Haynie 1974).

The uses of steel products and their distribution in the economy have been estimated by Fink, Buttner, and Boyd (1971). They isolated fourteen categories of products, representing a significant fraction of the total of steel products in use that might be exposed to the outdoor environment. They excluded products with a service life less than the expected maintenance interval of the surface and products, such as automobiles, whose surface maintenance is dictated by other considerations. They developed estimates of the fraction of the products that would be exposed, the fraction of the surface area exposed, and in a few instances the distribution of the exposed area among different land-use zones. They then calculated the in-place stock from annual production statistics and U.S. Internal Revenue Service (IRS) depreciation-lifetime guidelines, a procedure that probably results in an underestimate. Their estimates are shown in table 10-2.

The rate of corrosion of steel in polluted atmospheres has been studied both in the field and in the laboratory. Several dose-response functions have been suggested. Perhaps the most useful and reliable is the one proposed by Haynie and Upham (1974) from a field study of fifty-seven sites across the United States. Their results imply that, if the service life of an item at a specific site is limited by corrosion, the expected service lives (SL) at different levels (0 and 1) of sulfur dioxide would be related as

$$\frac{SL_1}{SL_0} = \exp 0.0055[(SO_2)_0 - (SO_2)_1] \qquad (10.1)$$

where the sulfur dioxide concentration (SO_2) is expressed in $\mu g/m^3$. Steels that are specifically alloyed to resist corrosion (high-copper steels, stainless steels, and "weathering" steels) would not behave as predicted by this relation. For some steels the rust layer actually acts to protect the metal from further corrosion.

Table 10-2
Steel Products Subject to Pollution Damage in 1970

Product Class	Protective Coating	Stock in Use (kg per capita)	Area Exposed (m^2 per capita)
Storage tanks	Paint	54	0.7
Bridges	Paint	83	1.0
Sheet metal	Galvanized	117	20.7[a]
Barwork, grills and gratings	Paint	14	0.5[b]
Structural steel	Paint	88	1.1[c]
Wires and cables	Galvanized	29	76.1m[d]
Fence fabric	Galvanized	18	1.8[e]
Fence posts	Galvanized	28	0.8
Power transformers	Paint	[f]	0.3
Street-lighting fixtures	Paint	4	0.4
Pole hardware	Galvanized	10	0.5[g]
Power-line towers	Galvanized	7	0.2

Source: Calculated from data in F.W. Fink, F.H. Buttner, and W.K. Boyd, *Technical-Economic Evaluation of Air Pollution Corrosion Costs on Metals in the U.S.,* PB 198 453 (Columbus, Ohio: Battelle Memorial Institute, 1971).
Note: Stock data are for portion exposed on one side to outdoor environment.
[a]Distribution estimated to be 30-percent industrial areas, 25-percent commercial, 25-percent residential, 20-percent rural.
[b]Distribution of 10-percent industrial, 5-percent commercial, 85-percent residential.
[c]Distribution of 60-percent industrial, 40-percent commercial.
[d]Length of exposed cable.
[e]Nominal area of two faces of fencing. Distribution is "predominantly" industrial.
[f]Not meaningful.
[g]Distribution of 45-percent urban, 55-percent rural.

Galvanized Zinc

Several dose-response functions have been suggested for zinc surfaces. Haynie (1980) reviewed the various studies and concluded that the differences in results were related to differing wind speeds at the experimental sites. He argues that, taken together, the various experiments support a relation for the corrosion rate (R, in μm/yr) of

$$R = [0.32 + F(U^{0.78})\,SO_2]\,RH/(100 - 0.86RH) \qquad (10.2)$$

where the sulfur-dioxide concentration is in $\mu g/mg^3$, relative humidity (RH) is expressed as a percentage, and the average wind speed (U) is in m/sec. The dimensionless factor, F, varies for different configurations of the material, from 0.0013 for large sheets, to 0.0019 for smaller sheets, and to

Estimating Materials Benefits

0.0028 for wires. This implies a corrosion rate about twice as fast on fence fabric and other wire surfaces, which has been observed in practice.

For most products in an unpolluted environment, the life of the galvanized layer will exceed the service life of the item. If the galvanized layer erodes before the end of the item's useful life, it will be necessary to tolerate a corroded surface for the final few years of economic life, to clean and paint the surface, or to replace the item sooner than otherwise required. Thus it is necessary to know both the normal useful life and the expected life of the zinc protective layer in order to calculate the change in the service life. If the service life is determined by the life of the zinc layer under both air-pollution conditions being investigated, then

$$\frac{SL_1}{SL_0} = \frac{0.32 + F(U^{0.78})(SO_2)_0}{0.32 + F(U^{0.78})(SO_2)_1} \qquad (10.3)$$

where the terms are as defined for equation 10.2.

Gillette (1975) has used similar equations to calculate the change in the service life of galvanized metals and thereby to estimate the aggregate corrosion damage. He presents data on the thickness of the protective zinc layer, the expected service life in unpolluted air, and the replacement cost for the more-important products.

Paint Dose-Response Functions

Paints are applied both to provide a decorative finish and to protect the underlying material from exposure to the weather. Factory-applied and cured coatings generally resist air-pollution erosion or discoloration damage at the concentrations usually encountered. In many applications they are rapidly replacing galvanizing as a protective coating because of their greater resistance. Oil-based exterior paints are subject to significant film erosion from sulfur dioxide and ozone. The observed erosion rate of most latex-based exterior paints is relatively low, but roughening of the surface by pollution increases soiling. A "frosting" that degrades the appearance of dark-colored latex paints also may be related to sulfur-dioxide air pollution. Approximately 70 percent of the exterior paint in place is latex-based.

Paint-film failure may be caused by a simple loss of the film thickness through erosion, by a breaking of the film due to poor adhesion or a loss of film flexibility, or by corrosion of the surface under the paint film. Some paints are intended to be self-cleaning by the formation and erosion of a surface layer (chalking). Air pollution can accelerate this process, thereby shortening the service life of the paint. Latex paints are designed to be porous, allowing moisture to move through the film. If a latex paint is applied directly to a metal surface, moisture and sulfur oxides can pass through the paint film to the metal and initiate corrosion. Oil-based paints will inevitably contain a few pores that also will permit the development of such underfilm corrosion. Where the life of the paint film is limited by under film corrosion on a steel surface, the service life of the paint coating could be expected to change according to equation 10.1 when ambient sulfur-dioxide concentrations change.

Film erosion of oil-based paints has been measured in several laboratory and field experiments. A dose-response function can be calculated from the field observations of Campbell et al. (1974) and air-quality data recorded at the same time and reported by the U.S. Environmental Protection Agency (1972). These data suggest an erosion rate (in μm/yr) of

$$R = 0.05 + 0.03 \, SO_2 \qquad (10.4)$$

where the sulfur-dioxide concentration is in $\mu g/m^3$. All the elevated-sulfur-dioxide sites in this field study were in an area of annual average relative humidity of approximately 65 percent. An adjustment of the dose-response function to wetter or drier climates would be necessary, with both terms in the dose-response function increasing at higher relative humidity. The data available do not permit the construction of a dose-response function for ozone, although the effect appears to be of approximately the same severity.

The service life of an oil-based paint film in an unpolluted environment is approximately eight to twelve years, although repainting often occurs earlier, either for aesthetic reasons or because of paint-film failure from other causes. The service life under polluted conditions will be determined by the erosion rate, calculated as described previously, and an effective film thickness of the paint of approximately $20\,\mu$m. The replacement cost of exterior paint can be estimated as at least $40 per capita, in 1978 U.S. dollars.[1]

Textiles Dose-Response Functions

Air-pollution damage to textiles includes both the accelerated fading of fabric dyes and the loss of strength in fabric fibers. Sulfur-dioxide exposure reduces the fiber strength in cotton and nylon fabrics. Nitrogen dioxide and ozone can accelerate the fading of dyes in various fabrics. In general, cotton fabrics, which make up approximately 35 percent of the fabrics in use, are the most sensitive to pollution damage. Fabrics made from cellulose (rayon and acetates) are intermediate in sensitivity to damage, but the dyes used with them are the most sensitive to pollutants. They constitute only 5 percent of the fabrics in use and seldom are used outdoors. Man-made fibers from hydrocarbons (nylon and polyester) and the dyes that are used with them are relatively less sensitive. Together with the polyester-cotton-blend fabrics, they account for more than half of the fabrics in use.

Numerous studies have evaluated the fading of various fabrics and dyes when exposed to different pollutants. However, most of the fading studies have been oriented more toward identifying the sensitive pollutant–dye–fabric combinations than toward measuring fading rates for different air-pollutant concentrations and times of exposure. Objective measures of fading that correlate to subjective judgments on the acceptability of the fabric for continued use have been described (Beloin 1973). Haynie, Spence, and Upham (1976) report a dose-response function for the fading of a vat-dyed cotton drapery fabric, in terms of these fading units. However, it is not clear that this or any other single dose-response function will be too useful since it only applies to one specific dye–cloth combination.

The textile industry has invested considerable time and money in determining which dyes are subject to pollution damage. These dyes can be avoided, but the alternative is often a dye that is more expensive to make and use and that requires a higher-quality fabric. These extra costs may add 2 to 4 percent to the cost of the finished fabric.[2] Because the cloth produced must be able to withstand the worst ambient-pollution conditions, the increased costs are shared by all consumers, including those in low-pollution areas.

There are several studies of the effects of pollutants on fabric strength. A dose-response function can be constructed from the data reported by Brysson et al. (1967). They exposed five cotton fabrics at seven sites in St. Louis and five sites in Chicago for a year. They concluded that the loss in breaking strength of the fibers was significantly correlated to the exposure to sulfur dioxide and not significantly correlated to suspended-particulate exposure, although the latter may have had some effect. Data from three of the sites in Chicago with reliable sulfur-dioxide measurements suggest an equation for a change in the service life of cotton fabric (when limited by a loss of fiber strength) with a change in pollution level of

$$\frac{SL_1}{SL_0} = \frac{21.27\,(SO_2)_0 + 7.11}{21.27\,(SO_2)_1 + 7.11} \qquad (10.5)$$

where the average sulfur-dioxide concentration is expressed in ppm. Equation 10.5 is most applicable to industrial fabrics (such as tarpaulins) that are exposed continuously to the outdoor environment.

Other Materials Damages

One of the most-dramatic effects of air pollution has been the deterioration of carved-stone buildings and statuary. If the item damaged has an important place in the culture, then the value lost may be very high. Sulfur oxides react chemically with some building stone to form a water-soluble compound that is then removed by rain. Damage from particulate matter may occur during attempts to clean the soiled surfaces. Intricate carvings may be destroyed by the removal of a relatively thin surface layer, even though structural walls without carved details may withstand substantial surface erosion without obvious ill effect. It is not clear that dose-response functions could be usefully applied in these diverse circumstances, even if they were available.

Air pollution also can damage electronic equipment. Corrosion may damage electrical contacts and insulators, or particulate matter may form current-leakage paths. Particulate matter has been observed to damage rotary switches and motors. The major economic loss from such damage has been found to be the loss of operating time when repairs are required (ITT Electro-Physics Laboratories 1971). No dose-response functions are available in the literature.

Ozone exposure will cause rubber and similar elastomers to become brittle and crack. This is observed even at natural background concentrations of ozone. To protect the rubber and extend its useful life, certain antioxidants and waxes are mixed with the rubber. The use of these special compounds adds approximately 1 percent to the cost of vehicle tires and 0.5 percent to the cost of other rubber products (Mueller and Stickney 1970). Waxes that will provide protection for three to five years in most areas have been observed to provide protection for one to two years in the higher ozone concentrations of southern California. Several studies of the service life of rubber products at varying ozone levels have been reported, but the relations they suggest are not consistent.

Valuation of Materials Damages

The economic loss from air-pollution damage to most materials is felt primarily in their more-frequent maintenance or earlier replacement, rather than in an interim decrease in the serviceability of the product. This is especially true when the damage is to a protective surface that has no other role.

The net present value (NPV) of the change in the timing of future replacement or repair from increased pollution is

$$NPV = \frac{P}{(1+\gamma)^n} + \frac{P}{(1+\gamma)^{2n}} + \frac{P}{(1+\gamma)^{3n}} + \cdots - \frac{P}{(1+\gamma)^m}$$

$$- \frac{P}{(1+\gamma)^{2m}} - \cdots$$

where P is the replacement price, n is the service life at the increased pollution concentration, m is the service life at the original level of pollution, and γ is the discount rate. These infinite series can be written as

$$NPV = P\left[\frac{1}{1 - (\frac{1}{1+\gamma})^n} - \frac{1}{1 - (\frac{1}{1+\gamma})^m}\right]$$

If we wish to express this as an equivalent annual cost, it is simply $(NPV)\gamma$. Then

$$\text{Annualized Cost} = P(\text{CR}_n^\gamma - \text{CR}_m^\gamma) \qquad (10.6)$$

where CR_n^γ is the capital-recovery factor for n years at the discount rate γ. In the first-order approximation the annualized cost is simply $P(1/n - 1/m)$. The error between the correct form and the approximation is only about 2 percent at commonly encountered values of n, m, and γ.

If we think of a large community where many people are carrying out maintenance activities each year, then $P(1/n - 1/m)$ is analogous to the change in price term, ΔP_a, defined in chapter 6. The errors of approximation associated with the demand and supply characteristics of the commodity described there also will apply here. For example, if the service life of an item increases, its effective annual cost decreases and, therefore, its use should increase. Thus estimates obtaned using equation 10.6 generally will overestimate the damage from increases pollution and underestimate the benefits from reduced pollution.

Summary

The principal economic effect of air-pollution damage to materials is the shortening of service life and maintenance intervals. Reliable dose-response functions have been reported for the corrosion of steel and zinc surfaces by sulfur dioxide. With knowledge of critical corrosion thicknesses, the change in service life can be estimated. Dose-response functions also are available for paint and cloth exposed to sulfur dioxide, but they are derived from very limited data. Functions for the fading of cloth probably are not useful since they apply to a single pollutant-dye-cloth combination that seldom will remain in production after the problem becomes apparent.

Notes

1. Huber (1978) reports that 30 million gallons of oil-based paint at a wholesale value of $205 million were produced in the United States in 1977. Use of the technique of Spence and Haynie (1972) to estimate an average useful life of 6.4 years and in-place value factor of 6.5 (to include the labor value in the application of the paint) yields an in-place replacement value of slightly less than $40 per capita. Fink, Buttner, and Boyd (1971) found an average painting cost of approximately $9.25/m^2 (assuming professional labor, in 1978 U.S. dollars). This would imply an in-place value factor greater than 20 and an in-place replacement value of more than $120 per capita.

2. Victor S. Salvin, University of North Carolina, Greensboro, personal communication, 1979.

11 Estimating Aesthetic Benefits

Although the economic loss from air-pollution damage to aesthetic values may be the most difficult to estimate, it actually may be larger than the losses from damage to vegetation and materials. Air-pollution damage to aesthetic values is caused primarily by odors associated with a variety of pollutants and by increased soiling and reduced visibility from suspended particulate matter.

This chapter presents dose-response functions for the rate of soiling and the change in visibility. Data also are presented for the monetary values associated with these aesthetic qualities. An alternative approach to the estimation of air-pollution damage to aesthetic values is the measurement of the loss of property value from air pollution. This will include soiling damages and other aesthetic losses that are associated with the location of the property. A brief review of four representative property-value studies concludes the chapter.

Odor Damages

The degree of an individual's annoyance with an odor will depend both on personal reactions to the character and intensity of the odor and on the frequency and duration of odor episodes. Techniques exist that can describe the intensity of an odor as a function of its concentration in the air and can characterize its relative pleasantness or unpleasantness (Dravnieks and O'Neill 1979). Community surveys have described the reaction of the public to specific odor situations (for example, Jonsson, Deane, and Sanders 1975). However, these two lines of research have not been brought together into a single model that could generate useful dose-response functions. The results to date indicate that a public-annoyance dose-response function must include socioeconomic and attitudinal variables. Willingness-to-pay surveys may prove to be the most direct way to evaluate the degree of community reaction to an odor.

Soiling Damages

Exposed materials will pick up soot and dirt from the air. Rain may wash off some dirt but may also deposit dirt. The character of the suspended particulate may vary significantly from city to city, being blacker and stickier in some areas.

Beloin and Haynie (1975) measured the rate of soiling of six common building materials over a two-year period in Birmingham, Alabama. They exposed wood siding painted with both latex and oil-based white paint, white asphalt shingles, concrete block, common brick, limestone, and window glass at five sites that experienced a gradient of suspended-particulate concentrations. Once a month the reflectance of the opaque surfaces was measured and was found to be directly related to a subjective appraisal of the cleanliness of the surface (with 100-percent reflectance equal to a clean mirror). The dirtiness of the window glass was measured in transmittance. The measurements of percentage of reflectance for the latex and oil-based paint samples could be expressed in a dose-response function of the form

$$A_0 + A_1\sqrt{(\text{TSP})}\, t,$$

where the total suspended particulate (TSP) is in $\mu g/m^3$ and time (t) is in years. The soiling of the shingles was best expressed in a similar function,

$$A_0 + A_1(\text{TSP})\, t,$$

with the same units. The estimated coefficients for these equations are reported in table 11-1. No dose-response functions were found that described adequately the soiling of the concrete block, brick, limestone, or window glass.

In an attempt to measure jointly the physical rate of soiling and personal reactions to it, surveys of household-cleaning operations have been made in three areas in the eastern United States. An early study by Ridker (1967) in Philadelphia, Pennsylvania found inconsistent relationships between cleaning frequencies and measured ambient concentrations of suspended particulate. There were substantial differences among his sample households in education, income, ethnic composition, and type of home-heating fuel; but the influence of these variables was not explored.

Michaelson and Tourin (1969) surveyed three towns with different pollution levels near Pittsburgh, Pennsylvania in 1960, and three towns, again with different pollution experiences, near Washington, D.C. in 1967. Although there were anomalies in the data for some of the cleaning operations, for most of the cleaning activities analyzed there were clear trends of increasing frequency of cleaning with increasing air pollution. For example, the frequency of repainting exterior wood siding increased with increases in suspended-particulate levels across all six towns studied. A dose-response function for the annual frequency of $-0.14 + 0.006$ TSP (in $\mu g/m^3$) is implied by the data they report. The data are for an upper-middle income

**Table 11-1
Estimated Coefficients for Beloin and Haynie Soiling Dose-Response Functions**

Material	A_0	A_1
Latex-painted wood siding	91	-1.74
Oil-base-painted wood siding	89	-0.96
Asphalt roofing shingles	42	-0.03

Source: Adapted from Norman J. Beloin and Fred H. Haynie, Soiling of building materials, *Journal of the Air Pollution Control Association* 25(1975):399-403.

segment, and thus this relation is not a general regression from socioeconomic and air-pollution data. Michaelson and Tourin do not report the socioeconomic data they collected nor do they report the statistical characteristics of their measured frequencies.

Booz, Allen, and Hamilton (1970) conducted a detailed study in the metropolitan Philadelphia area with the intent of producing statistically sound dose-response functions and expenditure data. They interviewed 1,500 households with respect to twenty-seven cleaning and maintenance activities. Their interview included a fifty-six-item attitudinal test, designed to establish the respondent's attitudes toward cleanliness. They found that the performance frequencies of eleven of the twenty-seven activities tended to increase with the suspended-particulate levels. Unfortunately, dose-response equations that included the attitudinal variables and socioeconomic data could not be constructed since most of these variables were highly collinear with air pollution.

Liu and Yu (1976) used the Booz, Allen, and Hamilton data to produce dose-response functions that take into account only the variations in suspended-particulate levels. They utilized these dose-response functions to predict the costs of soiling damage in 148 cities across the United States. These estimates, along with socioeconomic and climatic data from these cities, were then used to generate a damage function for soiling costs that includes socioeconomic and climate variables. This function may illustrate some relation between climate and local suspended particulate levels, but it cannot be considered a valid damage function for soiling.

Valuation of Soiling Damages

Watson and Jaksch (1978) have used data on cleaning frequencies and the cost of cleaning operations to derive estimates of the damage from soiling caused by air pollution. They argue that there is no substantial evidence that cleaning frequencies increase as pollution increases and that, therefore, the frequency of cleaning and expenditures on cleaning may be assumed to be constant. With some additional assumptions, discussed later on, about the supply of and demand for cleanliness, they derive expressions for the dollar value of damages from a change in particulate concentrations. Under their assumptions, their formula for the dollar value of damages is

$$\frac{E}{2}\left(\frac{\text{TSP}_0}{\text{TSP}_1} - \frac{\text{TSP}_1}{\text{TSP}_0}\right) \tag{11.1}$$

where E is the (constant) expenditure on cleaning. Using this equation and expenditure data collected in the Booz, Allen, Hamilton (1970) study, they estimate a marginal benefit from reduced soiling of approximately $3 per $\mu g/m^3$ TSP (in 1978 U.S. dollars) per household.

The dollar value of damages from soiling can be calculated under less-restrictive assumptions using equation 6.12. For a uniform shift of the supply function, the change in total consumers' and producers' surplus is

$$\Delta\text{CPS} = \frac{1}{2}\left[E_0 - E_1 + E_0\left(\frac{Q_1}{Q_0}\right) - E_1\left(\frac{Q_0}{Q_1}\right)\right] \tag{11.2}$$

$$- \frac{1}{2\xi_0}\left[E_0 - E_0\left(\frac{Q_1}{Q_0}\right)^2\right]$$

The Watson-Jaksch result can be obtained from equation 11.2 by imposing their assumptions, which can be expressed as $E_0 = E_1$, $\xi_0 = \infty$, and Q_j inversely proportional to TSP_j. Alternatively, the more general expression may be used directly. For example, if the perceived cleanliness of an external wall decreases proportionally to the reflectance as measured by Beloin and Haynie (1975), and the frequency of repainting is given by the expression obtained from the Michaelson and Tourin (1969) data cited previously, then the average level of cleanliness will be approximately constant for all relevant concentrations of total suspended particulate (that is, $Q_1/Q_0 \approx 1$). In this case, the value of soiling damages estimated by equation 11.2 is simply $E_0 - E_1$. This can be expressed in terms of the frequency of the cleaning operations as $E(1/n - 1/m)$.

The costs of several typical household cleaning operations are shown in table 11-2. The average replacement value of painted surfaces can be estimated at approximately $140 (1978 U.S. dollars) per capita.[1]

Table 11-2
Household-Cleaning-Operation Unit Costs

Operation	Average Cost[a]	With Imputed Costs[b]
Paint exterior walls	$410	$600
Paint exterior trim	150	210
Dry-clean draperies	120	140
Wash windows inside	2	12
Wash windows outside	3	17

Source: Calculated from data in Booz, Allen, and Hamilton, Inc. *Study to Determine Residential Soiling Costs of Particulate Air Pollution,* PB 205 807 (Washington, D.C.: Booz, Allen, and Hamilton, 1970).
Note: Adjusted to 1978 U.S. dollars.
[a]Does not include any labor costs for the fraction of homes that perform the task with their own labor.
[b]Assumes that household labor is valued at one-half the professional-labor rate.

Visibility Damages

The impairment of visibility through the atmosphere is caused primarily by very small particles that scatter and absorb light. Larger particles and colored gases, such as nitrogen dioxide, also reduce visibility, but to a lesser degree. Reduced visibility will be noticed both as a decrease in how far away objects can be seen clearly and as a change in the apparent color of the sky and of the objects viewed.

When a light ray strikes a particle in the air, it may be either absorbed or scattered (that is, deflected into a new direction). If light coming from an object is subject to scattering, less of it will reach the observer. The greater the distance between the object and the observer, the greater will be the number of particles in the way and the greater will be the reduction in the light. At the same time, other particles in the air will scatter light into the observer's sight path. The combination of these two actions will decrease both the contrast of the object relative to the horizon sky and the ease with which it can be observed. Since there is a slight difference in the way different colors of light are scattered, a haze also can cause a change in the apparent color of an object or a haze layer can itself appear to be colored.

The physical theory of light scattering and visibility has been thoroughly described by Middleton (1952). *Visibility,* or more precisely *meteorological range,* is formally defined as the distance at which a black object has a contrast of only 2 percent with the horizon sky. The perceived contrast (C) of an object is a function of the original contrast (C_0) between the object and the horizon and the distance (L) between the object and the observer. It is given by

$$C = C_0 \exp(-b_{ext}L),$$

where b_{ext} depends on the amount and kind of particles in the intervening air and is called the *extinction coefficient*. Thus b_{ext} is the fractional reduction in contrast per meter. Then the distance (in kilometers) where the contrast is reduced to 2 percent of C_0 is

$$L_V = \frac{3.9 \times 10^{-3}}{b_{ext}} \qquad (11.3)$$

The extinction coefficient is composed of four terms, for the scattering of light by (1) particles and (2) gas molecules and the absorption of light by (3) particles and (4) gases. In rural areas the absorption terms seldom will be more than 5 percent of the extinction coefficient, whereas in urban areas the absorption terms may account for as much as 40 percent of the total extinction (Weiss et al. 1979). The scattering coefficient (b_{scat}) of clean air is approximately 1.1×10^{-5} per meter. If there are no particles in the air and there is no absorption by gases, then $b_{ext} = b_{scat}$ and 1.1×10^{-5} can be divided into 3.9×10^{-3} to find that the natural light scatter of air molecules limits the visibility to about 350 km. If particles that do not absorb light are added to the clean air, visibility will decrease according to the sum of the scattering coefficients of the particles and the air. If the particle-scattering coefficient is, for example, 6×10^{-5} per meter, then the extinction coefficient is 7.1×10^{-5} per meter, and the visibility is approximately 55 km.

Fine particles with a radius between 0.1 and $1\mu m$ are significantly more effective in scattering light than the same mass of larger or smaller particles, because the wavelength of visible light is approximately $0.5\mu m$. Thus large particles, such as road dust, reduce visibility only when they are present in high concentrations. Sulfates and other particles formed from gases in the air are concentrated in the size range between 0.1 and $1\mu m$ and thus are among the most significant for light scattering.

Measurements by Waggoner and Weiss (1980) suggest a relationship between the total mass of suspended particulate matter less than about $2\mu m$ in diameter (M_f) and the scattering coefficient (expressed as 10^{-5} per meter) of

$$b_{scat} = 1.1 + 0.3\,M_f \qquad (11.4)$$

where the mass of fine particulate is in $\mu g/m^3$. It must be emphasized that it would be quite incorrect to use traditional measurements of total mass (that

Estimating Aesthetic Benefits

is, TSP rather than fine-particle mass) in this equation to generate estimates of the scattering coefficient.

If the change in the fine-particle mass is known, then equation 11.4 can be substituted into equation 11.3 to find the change in visibility that will occur:

$$\Delta L_V = - \frac{L_0 \Delta M_f}{(1300/L_0 + \Delta M_f)} \tag{11.5}$$

where L_0 is the initial visibility condition. It can be seen from this dose-response function that in relatively clean-air areas (L_0 large) a small increase in fine-particle mass will result in a significant decrease in visibility, whereas in a polluted area (L_0 small) it will take a substantial change in fine-particle mass to obtain a significant change in visibility.

Equation 11.5 describes the change in the visibility at a single time. It would be necessary to sum the daily changes over a year using an equation similar to equation 8.1 to obtain the annual average change. Figure 11-1 illustrates the annual average visibility predicted by equations 11.3 and 11.4 for a wide range of expected values of the annual average fine-particle concentration. The upper curve assumes there is no absorption ($b_{abs} = 0$), and the lower curve assumes the absorption is half as large as the particle scattering. This calculation does not take into account the reduction in visibility from natural meteorological conditions, so it will consistently overstate the actual visibility observed in all but very dry locations.

Detailed computer-based simulation models have been developed (for example, Latimer and Samuelsen 1978) to calculate the visibility impact of a single pollution source. These models consider the spatial relationships among the observer, the object viewed, and the pollution source, as well as various environmental conditions and varying assumptions about the rate of sulfate formation in the smoke plume. Ensor, Sparks, and Pilat (1973) have described a simplified model of visibility degradation that will occur downwind from a single source. This model will produce rough estimates with only a few calculations and a minimal amount of initial information. A similar model describing the visual appearance of the smoke plume has been detailed by Ensor and Pilat (1971).

Valuation of Visibility Damages

There have been a few attempts to estimate the value of visibility through surveys of the public's willingness to pay for clear air. Loehman et al. (1979) questioned residents of Tampa, Florida on the value of "haze in the air." They report a median bid of $1.95 (1978 U.S. dollars) for avoiding one

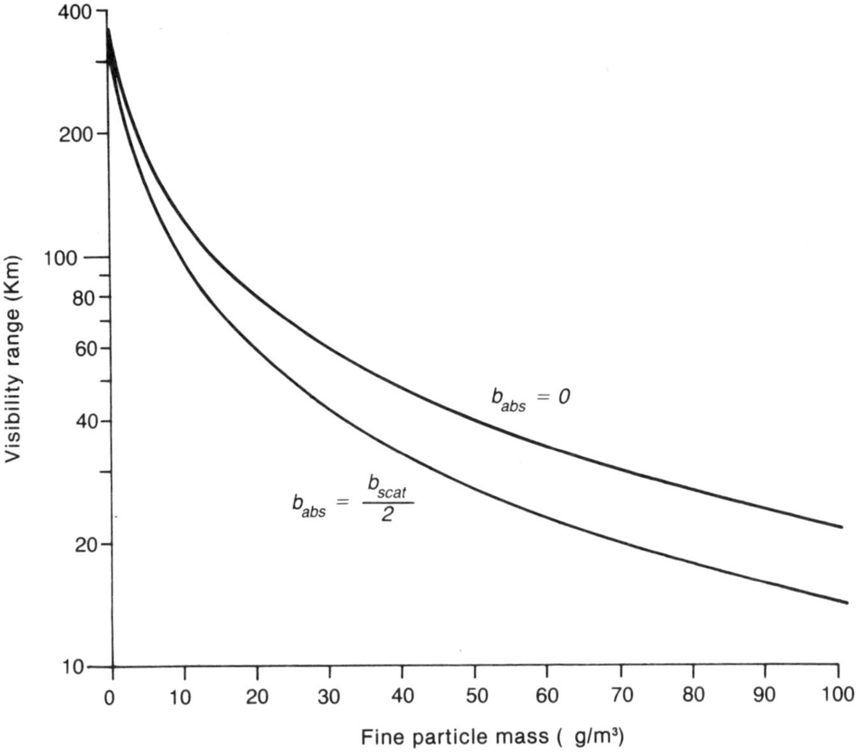

Note: Geometric standard deviation of fine particle mass concentration is assumed to be 2.0.

Figure 11-1. Visibility as a Function of Fine-Particle Concentrations

additional day of haze. A regression analysis of the bids indicated that willingness to pay increased with age and income and was greater for women than for men.

Other studies have attempted to define the visibility objectives more clearly by showing the survey respondents a set of photographs of local scenery with distinctly different visibility conditions. One study was conducted in Farmington, New Mexico (Rowe, d'Arge, and Brookshire 1980), which may have the greatest annual visibility in the United States. This visibility is threatened by the continuing construction of coal-fired power plants in the vicinity. Respondents were asked both what they would be willing to pay to prevent a degradation of their visibility and how much they would have to be paid to permit a degradation of their visibility. The bids obtained for three changes in visibility are shown in table 11-3. Almost two-thirds of the respondents refused to participate (or insisted on effectively infinite bids) in the compensation portion of the survey. Even with these respondents excluded, the average compensation bid is substantially higher. If expressed as dollars per month per kilometer change in visibility, the average willingness-to-pay is $0.10 and the average compensation bid is $0.89/mo/km. The income elasticity of the bids is estimated to be 0.4.

A study in Los Angeles, California (Brookshire et al. 1979) employed the same techniques to obtain estimates of willingness to pay to improve one of the worst urban visibility problems in the United States. The results for three changes in the average visibility are shown in table 11-3. The average bid is $0.24/mo/km. Because the visibility problem in Los Angeles also is associated with eye irritation and other overt health effects, a test was conducted to determine whether the bid for visibility improvement might include a component of willingness to pay to avoid the health effects. Half the respondents were first asked to bid to improve health and then asked what additional sum they would pay to improve visibility. For the other half, the order was reversed. In both instances the first bid tended to be much larger than the last bid. This may have reflected some sense of a budget constraint, although there is some evidence in their data of a possible interviewer bias being partly responsible for the difference. The average bid for visibility improvement with the visibility values asked first was $0.37/mo/km; with the health values first, it was $0.13/mo/km.

Property-Value Studies

A less-direct approach to estimating the value of aesthetic damages is the analysis of real-estate values. If air pollution decreases the usefulness or desirability of property in a way that is evident to both buyer and seller, then the sales price will reflect anticipated air-pollution damages. Pollution that normally cannot be detected by an individual should not create a price differential between two otherwise identical properties. Thus we would expect pollution price differentials to include losses from increased maintenance, the aesthetic damages of odors, soiling, ornamental-vegetation injury, and the acute health effect of eye irritation. Other health effects may be captured by property-value measures if they are perceived by the property owners and the buyers. Air-pollution damage that is fairly uniform across an urban area (such as visibility) would not be expected to create price differentials. Air-pollution losses incurred away from the home will not be captured by studies of residential-property values. Price differentials on agricultural property would reflect the economic loss to the crops. Care must be taken to avoid double counting when using property-value data in a benefit-cost analysis.

The sales price, or capital value, of a property (land and structures) will

Table 11-3
Mean Willingness to Pay for Visibility

Visibility Change	Farmington, New Mexico		Los Angeles, California:
	Pay to Prevent	Accept to Permit[a]	Pay to Improve
120 ↔ 80 km	$4.75	$24.47	
80 ↔ 40 km	3.53	46.63	
120 ↔ 40 km	6.54	71.44	
3 ↔ 19 km			$ 3.35
19 ↔ 45 km			6.93
3 ↔ 45 km			10.38

Source: Adapted from David S. Brookshire, Ralph C. D'Arge, William D. Schulze, and Mark A. Thayer, *Experiments in Valuing Non-Market Goods: A Case Study of Alternative Benefit Measures of Air Pollution Control in the South Coast Air Basin of Southern California,* Methods Development for Assessing Tradeoffs in Environmental Management, vol. 2, EPA 850/6-79-001b (U.S. Environmental Protection Agency, 1979) and Robert D. Rowe, Ralph d'Arge, and David S. Brookshire, An experiment on the economic value of visibility, *Journal of Environmental Economics and Management* 7(1980):1-19.
Note: Units are 1978 U.S. dollars per month.
[a] Infinite bids excluded.

represent the present value of the stream of all future services that the property is expected to provide. The stream of services will be the anticipated annual revenues from an agricultural or commercial property or the imputed rental value of housing services and neighborhood amenities for a residential property. Similarly, the pollution price differential represents the capital value of all the future damages from pollution.

In order for sales prices to adjust fully to reflect air-pollution damage, there must be complete (and costless) mobility of the owners and users of property. Then those who require or value cleaner air will be willing to pay the appropriate premium for properties in clean-air areas, and those who do not value clean air will move to the less-expensive polluted properties. Cultural, economic, and institutional barriers may slow the movement of purchasers between sites and prevent full equalization of net values. The existence of a substantial in-place housing stock that was created in response to the demands of a different time will make it difficult for a purchaser to find property with all the characteristics he desires. To the extent that full equalization fails, welfare gains and losses will be retained by the property owners and will not be captured by the property value.

Measurement of the pollution price differential has received considerable attention. The size of the price differences, and thus the ease and certainty with which they can be measured, depends on both the supply of and the demand for properties with both clean air and the other attributes desired by various groups (for example, groups with different average incomes). If the number of sites with good air quality is limited and the demand is high, then the pollution price differential will be large. Conversely, if there are many sites with clean air, then the size of the price differential will be small. Because property values anticipate future conditions, a general public belief that air quality will improve will result in the air-pollution price differentials disappearing faster than the pollution. Detailed reviews of the theory and practice of estimating air-pollution effects on property values have been published by Rubinfeld (1978) and Freeman (1979b). Four studies will be discussed briefly here as representative examples of different approaches to the problem.

Ridker. In one of the earliest studies of air-pollution effects on property values, Ridker (1967) analyzed the economic loss that occurred to residences in a St. Louis, Missouri neighborhood when a previously innocuous industrial plant was taken over by a different firm and began producing strong odors. Ridker collected sale prices on approximately 1,000 homes in the affected area and approximately 2,700 homes in a nearly identical area nearby, for the five years preceeding the conversion and for four years after. Using regression analysis, he constructed a model of year-to-year property-value changes in the two areas. His model showed a statistically significant price differential developing in the affected area after the problem with the industrial odors began. It stabilized at approximately 10 percent of the sale price. At the time the study was published, there had been no resolution of the problem despite corrective orders by St. Louis authorities.

Nelson. The most frequently used technique for obtaining pollution price differentials is to estimate a regression equation between pollution concentrations and census-tract averages of owner-estimated residential values (taken as a part of the decennial census). Studies have shown these owner-estimates to be reasonably close to sales experience. Nelson (1978) developed a regression model for 456 census tracts in the Washington, D.C. metropolitan area using 1970 census data and 1967 and 1968 pollution data. In his regressions he included average site characteristics such as lot area and number of rooms, socioeconomic characteristics such as racial character and crime rate, the distance to employment centers, and the concentrations of total suspended particulate (TSP) and ozone. Using a log-linear form, an equation was obtained that explained 88 percent of the variation in the property values and that had all the coefficients statistically significant. From this equation Nelson estimated the pollution price differential for each census tract. At the average of each of the variables, the price differential due to pollution was found to be $98 (in 1978 U.S. dollars) per $\mu g/m^3$ (annual average) of particulate plus $12 (in 1978 U.S. dollars) per $\mu g/m^3$ (summer average) of ozone.

Using the estimates of the differential for each census tract, Nelson constructed supply and demand curves for air quality, including such variables as family income and family size. From these equations he estimated that the price elasticity of demand for air quality was approximately -1.3, the elasticity of supply about 0.7, and the income elasticity of demand approximately 1.0.

Smith. Most pollution-price-differential analyses attempt to establish a relation to the overall property values. Smith (1978) argues that the premium for clean air is correctly associated with the land alone and must be normalized to the size and price of the lot that is purchased. This premium can then be written as a function of the land value, the lowest price paid for land in the urban area, the lot size, and the price elasticity of demand for land. Smith calculated these premiums for 300 transactions in Chicago during 1971 from mortgage-appraisal records for newly constructed housing. His regression equation included two measures of accessibility to employment centers, four socioeconomic variables, two land-use variables, and TSP concentrations for each site. All the coefficients were statistically significant. The pollution premium on the land was estimated at $84 (in 1978 U.S. dollars) per $\mu g/m^3$ (annual average) of particulate for the normalized lot. To determine the income effects on the premium, Smith repeated the regressions after breaking the data set into two groups according to family income. For families with annual income less than $25,000, the pollution premium was $43 (in 1978 U.S. dollars) per $\mu g/m^3$ of particulate, whereas for high-income families the premium was $107.

Harrison and Rubinfeld. An equation for the total amounts families are willing to pay for clean air was developed from property-value data for metropolitan Boston, Massachusetts by Harrison and Rubinfeld (1978a). They first estimated the price differentials in property values for residential housing in 506 census tracts, using a log-linear regression equation with five variables for neighborhood differences, two for building structure, ten for accessibility, and one for air pollution. Unlike previous studies that estimated air-pollution data by interpolating data among a few monitoring sites, Harrison and Rubinfeld used a mathematical air-quality model to generate values specific to each census tract. To obtain the willingness-to-pay curve, the change in the estimated pollution price differentials for a unit change in pollution was calculated for each census tract and entered into a regression equation with family-income and air-pollution variables. The resulting equation implies an income elasticity of demand for clean air of approximately 1.0.

Summary

The amount an individual is willing to pay to avoid air-pollution damage to aesthetic values can be expected to depend strongly on personal values and income.

Estimating Aesthetic Benefits

Studies of soiling have measured cleaning frequency in order to estimate the damages caused by air pollution, but this is an appropriate measure only if the quantity of cleanliness remains constant as the pollution levels change. Dose-response functions have been reported for the soiling of painted surfaces and asphalt shingles. A relationship has been demonstrated between air pollution and painting frequency, although the data base is very limited. Taken together, these studies suggest that the quantity of cleanliness may remain constant, although the evidence is minimal.

The observed visibility is directly related to the scattering coefficient of the particles in the air. Less directly, it is approximately related to the mass of fine particles but not to the mass of total suspended particulate, since the large particles (which have most of the mass) are not as effective at scattering light. Economic studies have used surveys to estimate the public's willingness to pay for good visibility. They suggest a minimum value of $0.10/km per month for a change in visibility.

Econometric analysis of property values has revealed that people are willing to pay significant premiums for the perceived benefits of living in cleaner air. Several studies have found a household value of approximately $100 per $\mu g/m^3$ (annual average) of total suspended particulate.

Note

1. Huber (1978) reports that 100 million gallons of latex paint at a wholesale value of $525 million were produced in the United States in 1977. As described in chapter 10, this yields an in-place replacement value of approximately $100 per capita. The total in-place value for both latex and oil-based paint is thus $140.

12 Procedures for Evaluation of Control Costs

Most programs for the reduction of air pollution rely either on reducing the amount of the polluting substance that enters the production process (for example, taking the sulfur out of coal or the lead out of gasoline) or on removing pollutants from the gases released to the atmosphere. The latter most often is accomplished by "tail-end" control equipment designed to remove a specific pollutant. Sometimes the production process can be modified to reduce the amount of the pollutant released. In a few instances other techniques may be appropriate, such as increasing the height of a stack to dilute the pollutant to a safe concentration before it reaches ground level.

This chapter describes the procedures most commonly used for developing estimates of air-pollution-control costs.[1] Because the major source of variation in actual cost estimates is the choice of items to be included, rather than the estimated costs of individual components, the chapter includes a detailed listing and discussion of the cost elements that should be included in an estimate. The procedures for evaluating the economic impact of control expenditures on the firm are also discussed. Appendix B presents illustrations and brief descriptions of the common types of tail-end air-pollution-control equipment.

Types of Cost Estimates

Most cost estimates are based on relationships between the cost and the size of previous installations. The accuracy of estimates obtained in this way depends primarily on the choice of the size variable. The total production capacity of the plant will be only approximately related to the cost of the control equipment. The volume of exhaust gas treated provides a more appropriate measure of size but will not be an entirely accurate variable because other characteristics of the gas and the pollutant also will be important in determining the exact design of the required control equipment.

An estimate based on gross plant variables, such as total production capacity, is termed an *order-of-magnitude estimate* and generally will be accurate to within a factor of two. Given the necessary information, a person familiar with the air-pollution cost-estimation literature can make a responsible order-of-magnitude estimate in less than a day.

A *study estimate* is potentially more accurate because it is based on separate estimates of the costs of individual components of the pollution-control system. Development of a study estimate requires knowledge of the specific sources of pollutant emissions; a schematic design of the control system; approximate sizes for the components; the required control efficiency; and any unusual aspects of the plant, site, or pollutants. A study estimate can be expected to be accurate to ±30 percent. Achievement of that accuracy often will require a site visit (for a retrofit installation) and cooperation from the source operator. Preparation of a conscientious study estimate may take from several days to more than a month.

More-accurate estimates of cost are seldom available for use in a benefit-cost analysis. In order to make a *preliminary estimate,* with an expected accuracy of ±20 percent, it is necessary to have preliminary site-specific data such as the planned location of the equipment on the site; actual fan, motor, utility, and instrumentation requirements; and preliminary structural design of the control equipment. Development of such an estimate may cost as much as 1 percent of the total project cost. A *definitive* or *detailed estimate,* accurate to ±5 to ±10 percent, must be based on actual engineering drawings and suppliers' prices for specific items. These are the estimates on which bids are requested or made. They may cost 3 to 5 percent of the total project cost.

Cost-estimation data are often prepared in terms of *purchase cost*, which is the cost of the principal piece of control equipment delivered to the site but not installed (also called *flange-to-flange cost*). Some data are presented in terms of the *installed cost,* which should include the cost of interconnection and integration of the control equipment with the source and the cost of auxiliary equipment, but often does not. The installed cost is assumed to be for an unobstructed site (also called a "grass-roots" or "greenfields" site). The *total cost* will include these costs as well as indirect (that is, overhead) costs.

The most common procedure for converting from an estimated equipment-purchase cost or installed cost to a total cost is to multiply the equipment cost by a single number that accounts for all the extra costs involved. Average-total-cost multipliers have been suggested for different types of process equipment and for specific types of air-pollution-control equipment.

Studies of actual project costs have established approximate values for multipliers for installation, auxiliary equipment, and indirect costs as a percentage of the equipment-purchase cost for an "average" project (for example, Weaver and Bauman 1973). These may be summed to obtain a single multiplier, or they may be applied separately. If each cost item is con-

Evaluation of Control Costs

sidered separately, then the average values can be adjusted for the known circumstances of the particular project being analyzed. Tables are available that indicate a range of adjustments for commonly encountered variations (for example, Neveril 1978). After the adjustments are made, the terms may be summed and the total used as a single multiplier to convert the purchase cost or installed cost to a total cost.

An additional refinement of this technique divides the installation, auxiliary equipment, and indirect costs into a labor portion and a materials portion to permit adjustments reflecting local wage rates or unusual circumstances. Labor-materials ratios for many operations are reported by Guthrie (1970) and Blecker and Cadman (1973). However, the data necessary to utilize this technique fully seldom are available for a benefit-cost-analysis estimate.

Defining Control Requirements

Cost functions are available (see, for example, appendix C) that permit an order-of-magnitude estimate of the total cost of air-pollution abatement for many facilities with no more information than the total production capacity and, in some cases, the specific type of process in use. However, if this type of estimate is not sufficiently accurate for the benefit-cost analysis being conducted, specific information about the pollution source and the required control will be needed.

The current emission rates of air pollutants and the principal characteristics of the exhaust gases (location, height, and exit diameter of the stack, and volumetric flow rate and temperature of the gas) generally are catalogued in an "emission inventory" maintained by the local air-pollution-control authorities. If emission rates are not available either in the emission inventory or from the operator of the source (or proposed source), they can be estimated from emission factors (Anderson 1973; U.S. Environmental Protection Agency 1980b), which give the emissions that would occur without any air-pollution controls per unit of product produced. Emissions from motor vehicles are calculated similarly from emission factors (U.S. Environmental Protection Agency 1978e) and traffic volumes.

Future emission rates will be determined either by a specific regulation that establishes a permitted rate or by calculations from equations such as 7.1 or 7.2. Either way, the difference between the uncontrolled or existing rate and the projected rate will define the control efficiency that must be achieved by the proposed control program.

In order to prepare even a schematic design of a control program, it is important to have a basic understanding of the production process, the specific sources of pollutants within the process, and the appropriate control techniques to be used. There are several excellent sources of information. General descriptions of the major pollutant sources and the typical controls used can be found in the *Control Techniques* documents issued by the U.S. Environmental Protection Agency for each of the major pollutants (that is, U.S. Environmental Protection Agency 1969, 1977b, 1978b, 1978c, 1979b, 1980c). More-detailed descriptions are available for industrial processes for which U.S. New Source Performance Standards have been proposed (for example, U.S. Environmental Protection Agency 1974a, 1978d). These reports are particularly useful in describing how the production process can be modified to reduce pollutant emissions. In most cases these reports also contain an example of cost analysis and references to other sources of cost data. A useful source of information on the many industries that process and compound basic raw materials and chemicals is the comprehensive *Industrial Process Profiles for Environmental Use* (U.S. Environmental Protection Agency 1977d). This twenty-six-volume document contains information on many pollutant sources not described elsewhere. Finally, Neveril (1978) present summary tables of more than fifty sources, the controls commonly used, and typical values for many of the variables in equipment-design calculations.

Because process analysis and development of a schematic control program can take a significant amount of time and labor, cost estimation for a large number of similar facilities often is simplified by defining a few "model plants" with average characteristics and calculating study estimates for the model plants. The total costs then may be computed by scaling the costs calculated for the model plants to the production capacity of the actual plants.

Items Included in Cost Estimates

A cost estimate has little meaning unless there is a clear definition of what it includes. Despite efforts by professional organizations to standardize terminology, purchase costs reported in the literature sometimes include auxiliary equipment and differing degrees of delivery and installation. Installed cost may include only the construction of the control equipment on an unobstructed site, or it may include all the costs of bringing the equipment into operation (that is, all capital costs except the purchaser's expenditures for engineering design and supervision and production penalties during installation).

Table 12-1 lists the major categories of costs that will be encountered in implementing most pollution-control programs for a stationary source. Although it is specifically applicable to industrial sources, table 12-1 should provide a guide to the costs associated with control programs for other sources, such as transportation-control plans.

Evaluation of Control Costs

Table 12-1
Expenditures Included in Cost Analysis

Capital Costs

Direct costs

Control equipment purchase cost.
Auxiliary equipment, including exhaust hoods, ducting, fans, pumps, piping, conveyors, stacks, storage tanks, waste-disposal or pollution-control facilities, and so on.
Freight charges.
Site preparation, including demolition, construction of access roads and loading facilities, protective enclosures, lighting, and so on.
Installation, including foundations, erection, supporting structures, enclosures or weather protection, insulation and painting, connection of utilities, process integration, and so on.
Instrumentation, including process and emission monitoring, control panels, and so on.
Auxiliary buildings, including materials storage, employee facilities, modifications in existing structures, and so on.
Spare-parts inventory.
Working capital.
Ambient monitoring network.
Land.

Indirect costs

Engineering design and supervision, including preliminary source testing, pilot-plant studies, consultants, initial design, bid procedures, securing air-quality permits, administrative overhead, and so on.
Construction expense, including construction liaison, securing local building permits, insurance, interference with normal plant activities, and so on.
Contractor's fee and overhead.
Production penalties, including loss of normal production during construction or startup.
Startup, includes expenses of initial shakedown operation of control equipment and certification source testing.

Operating Costs

Operating labor, including supervision.
Overhead, including administration, payroll, occupational health and safety, and so on.
Utilities and consumable materials, including water, power, steam, oil or gas, limestone, and so on.
Maintenance labor, including supervision.
Maintenance parts, including replenishment of spare-parts inventory and major planned maintenance, such as catalyst or bag replacement.
Waste disposal, including charges for landfill, ponding, water clarification, and so on.
Production credits or penalties, including changes in plant productivity caused by operation of control equipment.
Product or energy recovery or byproduct credits, net of sales expenses if sold externally.
Operation and maintenance of ambient monitoring equipment.

Secondary Costs

Changes in the prices of materials, energy, byproducts, and so on, caused by the increased demand or supply created by the control program.
Regulatory, monitoring, and enforcement costs of the air-pollution-control agency.

Three costs often included in statements of control cost are conspicuously absent from table 12-1. Taxes are not included because in the type of benefit-cost analysis addressed in this book only social costs are significant. Taxes represent only transfers from the firms to the public, through the government, rather than true social costs.[2] Contingency reserves, which ordinarily would be included to cover expenses caused by unforeseen site problems, delays, or engineering errors, are more correctly handled as an element of uncertainty, as discussed in chapter 5. If an expenditure of this type is considered to be inevitable, it should be included in the cost item where it is expected to occur. Also excluded are such items as capital charges, depreciation allowances, and lost-opportunity costs. The costs associated with alternative investment opportunities already have been taken into account in the construction of the net-social-benefit formula 4.35. Including an additional cost of capital in the annual operating costs would distort the analysis. Similarly, a depreciation allowance is intended to provide funds for replacement of the facility at the end of its useful life. If it is desired to extend the analysis across more than one lifetime of the control equipment, replacement expenditures should be accounted for in the years in which they actually occur, rather than through the artificial bookkeeping entries of depreciation.

Included in table 12-1, but often omitted in practice, are two secondary effects. If the pollution-control program increases demand for energy or for a raw material consumed in the control process to the extent that prices change, then the loss in consumer surplus resulting from the increased price and reduced consumption in all other sectors must be included as a cost. Alternatively, if the control program results in the recovery of enough product or energy or the production of sufficient byproduct to cause lower prices, then the net gain in the surplus must be counted as a credit against the cost of the control program.

In making a cost estimate it is also important to include the costs incurred by the government agency in implementing and enforcing the control program. Depending on the specific program under analysis, these costs may result from efforts as simple as observing a compliance test or as complex as the development and execution of a state implementation plan.

Sources of Variability in Estimates

Substantial differences between two cost estimates are likely to reflect variations in the basic assumptions of the analysis, or the omission of specific operations, equipment, or facilities, rather than a gross error in the estimated purchase cost of a particular piece of equipment. Bloom et al. (1978) found that for similar installations of control equipment at electric-power plants, the major differences in cost estimates were caused by varying definitions of the project scope (that is, which costs were assumed to be a part of the control program) and to differing multipliers used to account for the costs that were not directly estimated.

The project scope should include all costs (and only those costs) that would be incurred only if the air-pollution-control project were undertaken. Deciding which costs to include can be difficult if the project is designed to solve several problems at once, such as occupational exposures to the pollutant and process modernization. However, costs clearly attributable to the control project, such as the cost of disposal of the collected wastes, should not be left out of the analysis simply because they have become part of another control program. An alternative approach is to modify the scope of the benefits estimates to include all the benefits attributable to the project. This may be particularly appropriate for transportation-control programs such as traffic-flow improvements, which also provide safety or fuel-economy benefits and where no one motive would have been sufficient to initiate the project.

Bloom et al. also found differences in project scope to include varying definitions of the required control technology. An oversized, high-reliability design of the control equipment might significantly increase the cost by including redundant equipment modules or large safety factors. McGlammery et al. (1975) have illustrated the problem by listing a set of reasonable changes in a control project that doubled the costs. Their possible changes included additional instrumentation and equipment for increased reliability, closed-loop waste-water control, bypass ducts and dampers, and cost escalation due to project delay.

The differences in the total-cost multipliers used by various analysts may reflect the varying circumstances in the projects they are drawn from or varying degrees of conservatism in their estimating procedures. Bloom et al. found the most important differences associated with site preparation and retrofit. The costs of installing the same control equipment in an existing facility were 25 to 50 percent greater than in new construction, depending on the age and size of the plant. Additional data on retrofit costs are reported by Devitt et al. (1976). Other site-specific variations include such problems as geographic differences in raw-materials composition and high costs resulting from the isolation of the site.

In a systematic comparison of the different cost-estimating methodologies used by fifteen agencies, companies, and consulting firms, El-Sawy, Leigh, and Trehan (1979) found that the greatest differences resulted from variations in the treatment of inflation and depreciation and in the distribution of the capital-investment payments over the construction years. The observed problems with inflation can be avoided by carefully treating inflation in a perfectly symmetric way. As discussed in chapter 4, if the benefits and costs are measured in constant dollars, then real discount rates must be used. However, if nominal interest rates are used, then benefits and costs in future years must be inflated using anticipated inflation rates. As discussed previously, charges for depreciation allowances and interest charges during construction should not be included, and expenditures during construction should be charged in the year in which they actually occur.

The correct lifetime for a project should be the expected effective life of the equipment. Neveril (1978) suggests a range of lifetimes for many types of control equipment. Lifetimes for process equipment can be estimated from the Asset Depreciation Ranges of the U.S. Internal Revenue Service (U.S. Department of Treasury 1979). These guidelines give anticipated lifetimes for process equipment in all major industry groups. As a general rule, a project lifetime of fifteen years is a reasonable choice, particularly if a large number of different components with different lifetimes are involved. If control equipment is purchased before the beginning of the analysis period or if it retains a useful life at the end of the period, both an initial value and a terminal value should be included in the analysis.

Price Adjustments to Cost Estimates

The purchase cost of a specific piece of air-pollution-control equipment will change over time because of changes in the prices of materials and labor and changes in design resulting from changes in the air-pollution-control requirements or changes in technology or manufacturing techniques. Changes in prices can be accounted for with a price index. Changes in costs caused by changes in design or technology must be estimated directly.

Inflationary price escalation has been a problem for analysts for long enough that most are now careful to report the year (and often even the month) on which their data are based. With this information and a price-index series, their reported equipment costs can be updated by multiplying the original estimate by the ratio of the current value of the index to its value in the year the original estimate was made.

A number of indexes are available for price adjustments in the United States. The ones most commonly used are listed in table 12-2. The Producers' Price Index is based on the prices of raw materials, intermediate goods, and finished goods at wholesale. The other indexes presented here are constructed from a weighted average of wages for several classes of labor and the prices of several types of materials. Differences in the year-to-year change in the indexes are caused by the different weightings assigned to the labor and materials components. The components and weightings of the M & S Index are described by Stevens (1947) and the Chemical Engineering (CE) Plant Index by Arnold and Chilton (1963). Current values of the

Table 12-2
Price Indexes for Air-Pollution-Control Projects

Year	M & S Chemical Industries	CE Plant Index	BEA Air-Pollution Equipment[a]	Producers' Price Index[b]	M & S Chemical Industries[c]	CE Plant Index[c]
1967	262.9	109.7	—	84.8	79.2	80.0
1968	273.1	113.7	—	86.9	82.3	82.9
1969	285.0	119.0	—	89.9	85.8	86.7
1970	303.3	125.7	—	93.3	91.4	91.6
1971	321.3	132.2	—	96.7	96.8	96.4
1972	332.0	137.2	100.0	100.0	100.0	100.0
1973	344.1	144.1	105.4	106.8	103.6	105.0
1974	398.4	165.4	122.8	130.4	120.0	120.5
1975	444.3	182.4	138.8	145.5	133.8	132.9
1976	472.1	192.1	147.3	154.7	142.2	140.0
1977	505.4	204.1	156.9	165.5	152.5	148.8
1978	545.3	218.8	166.3	177.6	164.2	159.5
1979	599.4	238.7	—	200.6	180.5	174.0

Source: Producer Price Index: U.S. Bureau of Labor Statistics; M & S and CE: Economic indicators, in each issue of *Chemical Engineering*; BEA: Calculated from Gary L. Rutledge, Pollution abatement and control expenditures in constant and current dollars, 1972-77, *Survey of Current Business*, 59, no. 2 (1979):13-20; and Gary L. Rutledge and Betsy O'Connor, Capital expenditures by business for pollution abatement, 1977, 1978, and planned 1979, *Survey of Current Business* 59, no. 6 (1979):20-22.

[a]The BEA index has not been calculated for years prior to 1972. The value for 1979 is not yet available.
[b]Index for all nonfood commodities. Renormalized to base year of 1972 = 100 for comparison.
[c]Renormalized to base year of 1972 = 100 for comparison.

M & S and CE Plant Index are published each month in *Chemical Engineering* magazine.

The M & S Index is primarily oriented toward process equipment, whereas the CE Plant Index includes components for buildings, engineering, and supervisory labor. Both indexes include an allowance for productivity gains by labor over time. Neither index is increasing quite as fast as the price deflator constructed specifically for air-pollution-control equipment by the U.S. Bureau of Economic Analysis (BEA).

Neveril (1978) recommends using one component of the CE index, the Fabricated Equipment Index, and various producers' price indexes of the U.S. Bureau of Labor Statistics to construct specific adjustments for individual parts of the purchase-cost estimate. He describes the components to be used and reports historic data for each component.

Adjusting costs calculated from relationships specific to U.S. installations to prices appropriate for other countries involves more than simply dividing by the exchange rate. Different prices and productivity of labor will result in a different mix of labor and materials in the final cost as well as different wage rates. Some suggestions for making such adjustments are given by Yen (1972) and Miller (1979). Construction-price indexes for a number of countries (in terms of their own currencies) are published in each issue of *Engineering Costs and Production Economics* (formerly *Engineering and Process Economics*). The construction of these indexes is described by Cran (1976).

Financial Analysis of Control Costs

In addition to estimating the economic costs of a proposed air-pollution-control policy, it may also be necessary to estimate the financial impact of the policy on a specific plant. The plant may meet the additional costs of installing and operating the equipment either by raising the prices of its products or by accepting reduced profits, but its ability to do either is limited.

Three tests have been used to determine the ability of a plant to absorb cost increases. First, market studies can evaluate the ability of purchasers to substitute other suppliers or products as the price increases. For some products, particularly industrial raw materials and fuels, this will be substantially determined by transportation costs.

Second, difficulties that the plant may face in obtaining financing for the air-pollution-control program may be estimated from a calculation of the debt-service-coverage ratio. This is the ratio of the before-tax profits to the annual payments on long-term obligations. As this ratio moves down from 2.0 to 1.5 and lower, financing new debt becomes increasingly difficult.

Evaluation of Control Costs

Third, changes in the profitability of the plant may be estimated from changes either in the return-on-investment ratio or in the discounted cash flow. The return-on-investment ratio is the ratio of the net profits to the asset value of the plant. This ratio may be compared to average returns for the industry or to the cost of new capital for the plant. The discounted cash flow measures the present value of future net cash flows from the continued operation of the plant. This test is used to determine whether it would be more profitable to close a plant in the face of a required pollution-control expenditure.

These tests are described more fully in U.S. Environmental Protection Agency (1979e). This reference also provides detailed schedules for calculating the necessary data. A more detailed, and more technical, discussion is given by Merrett and Sykes (1973).

Summary

The cost of control equipment most often is estimated by comparison to the known costs of previous installations. The quickest way to make an estimate, but also the least accurate, is to relate control-equipment costs to a gross plant variable, such as total production capacity or the amount of exhaust gas to be treated. Greater accuracy can be obtained by identifying the actual control requirements and developing capital-cost data for the necessary equipment from published equations or tables. Estimates can be developed in similar ways for installation, operating, and indirect costs.

The major source of variation in reported expenditures has been the items included in the cost analysis. All expenditures that would not have occurred if the control program had not been undertaken should be included. However, taxes, contingency reserves, and capital and depreciation charges should not be counted as costs.

Although it is not directly a part of benefit-cost analysis, decision makers often request an analysis of the financial impact of a control program on the firm. Several techniques are available for presenting this information.

Notes

1. Books recommended for additional information on cost-estimation procedures include Peters and Timmerhaus (1979), a widely used textbook that contains both introductory materials and detailed discussions of cost estimation for several processes, primarily from the chemicals industry. Weaver and Bauman (1973) provide a brief but comprehensive review and an extensive glossary of terms. Park (1973) and Woods (1975) contain well-

written discussions of general cost-analysis techniques and problems. Uhl (1979) describes cost-analysis techniques and data sources with special reference to pollution-control equipment.

2. However, as discussed in chapter 6, taxes may be relevant in calculating the value of large changes in consumption.

13 Data for Estimating Control Costs

The cost of implementing an air-pollution-control policy generally can be estimated with reasonable accuracy from published data for similar, previous programs. Cost functions have been developed for a wide variety of stationary sources and their control equipment, but only a limited amount of cost information is available for the control of pollution from motor vehicles.

Benefit-cost analyses most often will rely on either an *order-of-magnitude* cost estimate, which generally will be accurate to within a factor of two, or a *study estimate,* which can be accurate to ± 30 percent. The order-of-magnitude estimate is based on gross plant variables, such as production capacity or the volumetric flow of the gases to be controlled. For a study estimate it is necessary at least to calculate the approximate size of the control device that will be needed to achieve the required control efficiency; to identify the additional equipment required (such as ducting, fans, gas preconditioning equipment, storage tanks, and so on); and to identify any unusual circumstances with respect to implementation of the control program at the facility being examined.

This chapter reviews the available information for making control-cost estimates. Data are presented for use in order-of-magnitude estimates of costs for the more common types of air-pollution-control equipment. Cost functions are reported for flue-gas-desulfurization equipment at electric-power plants. Cost functions for other industrial sources are described in this chapter and are reported in appendix C. The diversity of air-pollution sources and of methods of controlling them makes it impossible to present here the necessary cost data for study estimates. Instead, references to sources of additional information are given. The chapter concludes with a brief discussion of the costs of controlling transportation pollutants.

Cost Functions for Order-Of-Magnitude Estimates

Order-of-magnitude cost estimates are based on empirical functions relating the cost and the size of air pollution control installations. In many studies, the preferred functional form for the cost function is

$$\text{Cost at New Plant} = (\text{Cost at Old Plant})\left(\frac{\text{Size at New Plant}}{\text{Size at Old Plant}}\right)^{\beta} \quad (13.1)$$

The exponent β is referred to as the *scaling factor*. If the value of β is 1.0, cost increases proportionally with the size of the installation. The estimated value of β in actual cost functions is usually less than 1.0, implying that cost increases more slowly than size. When sufficient data are not available on previous installations to permit estimation of β, it is often assumed to be equal to 0.6.[1] In a few instances, the value of β may be greater than 1.0, so that cost increases more rapidly than size. Bloom et al. (1978) observed this for air-pollution-control systems installed on existing electric-power plants, where the increased size led to greatly increased difficulties.

If data are available for a similar installation at a similar plant, then the scaling factor can be used to adjust the known value to the projected size of the needed control equipment at the new facility, with the accuracy of the result depending on how important the chosen size variable is in determining the cost of the equipment. Table 13-1 gives scaling factors that have been suggested by several authors. In most instances the size variable is the volume of exhaust gas that is received by the control device. The table also gives a range of values for the variable where the relation is believed to be valid. Some of the suggested factors in table 13-1 are inconsistent. This can be caused by using too diverse a range of data when estimating the scaling factor.

Table 13-1
Scaling Factors for the Purchase Cost of Air-Pollution-Control Equipment

Type	Range/Variable	Exponent	Source
Adsorber	10–150 cfm at 100 °F	0.32	Woods (1975)[a]
	150–1500 cfm at 100 °F	0.67	Woods (1975)
Cyclone	2–(7 × 10^3) acfm	0.61	Weaver and Bauman (1973)
	10^3–(8 × 10^4) scfm	0.56	Woods (1975)
Cyclone, multi-	10^3–(1.5 × 10^5) scfm	0.66	Woods (1975)
Electrostatic precipitator	10^4–(4 × 10^4) acfm	0.68	Weaver and Bauman (1973)
	10^3–(8 × 10^4) cfm at 100 °F	0.39	Woods (1975)
	(8 × 10^4)–10^6 cfm at 100 °F	0.81	Weaver and Bauman (1972)
Fabric filter	2–10^4 acfm	0.70	Weaver and Bauman (1973)
	(5 × 10^3)–10^5 acfm	0.80	Roeck and Dennis (1979)
	10^5–(4 × 10^5) acfm	0.72	Roeck and Dennis (1979)
	(2.5 × 10^5)–(2.5 × 10^6) acfm	0.62	Roeck and Dennis (1979)

Data for Estimating Control Costs

Table 13-1 continued

Type	Range/Variable	Exponent	Source
Fabric filter, shaker	10^3–(5×10^4) scfm	0.79	Woods (1975)
Fabric filter, reverse-jet	(5×10^3)–(6×10^4) scfm	0.71	Woods (1975)
Fuel-desulfurization unit	(2×10^3)–(4×10^4) bbl/day	0.65	Guthrie (1970)
Mist eliminator	(8×10^3)–(2×10^5) scfm	0.72	Woods (1975)
Packed tower, without fans	(4×10^3)–(3×10^4) scfm	0.68	Woods (1975)
Scrubber, impingement, without fans	10^3–(7×10^4) scfm	0.68	Woods (1975)
Scrubber, venturi, without fans	10^3–10^5 scfm	0.50	Woods (1975)
Scrubber, spray, without fans	(5×10^2)–10^5 scfm	0.72	Woods (1975)
Sulfuric-acid plant	(2×10^4)–(5×10^5) tons/year	0.65	Guthrie (1970)

Note: acfm = actual cubic feet per minute; scfm = standard (at 70 °F) cubic feet per minute; bbl = barrell.
[a]Donald R. Woods *Financial Decisions in the Process Industries* (Englewood Cliffs, N.J.: Prentice-Hall, 1975) also reports base costs and sizes, to complete equation 13.1.

An alternative form of equation 13.1 can be obtained by combining the data for the existing facilities into a single term,

$$\text{Cost at New Plant} = \alpha(\text{Size at New Plant})^\beta \quad (13.2)$$

where α is determined by the costs and sizes of the previous installations. With values for α and β, estimates can be made without the need to find a separate purchase cost at a similar facility.

A large number of such cost functions have been developed for use in two studies of the effects of pollution-control expenditures on the U.S. economy. One of the studies is the U.S. Environmental Protection Agency's (1979c) "Cost of Clean Air" project, which estimates future national expenditures due to air-pollution-control legislation. The other is the Strategic Environmental Assessment System (SEAS) model (House 1977), which is currently supported by the U.S. Department of Energy and the U.S. Environmental Protection Agency. This model has been developed to evaluate the effects of changes in pollution regulations not only on pollution-control expenditures but also on the supply of and demand for energy and on secondary effects on the economy. Over a hundred order-of-magnitude cost functions are used in these studies; these are described and listed in appendix C.

Control-Equipment Cost Functions

If enough information is available on the exhaust gases to make it possible to select a control technology, and if the volumetric flow to the control equipment is known, then an order-of-magnitude estimate can be calculated to determine the purchase cost of the principal types of control equipment, using equations developed by Kennedy et al. (1973) from data presented by Edmiston and Bunyard (1970). These equations are listed in table 13-2. Also listed in table 13-2 are operating and maintenance (O & M) cost equations for the "typical" conditions reported by Edmiston and Bunyard. They also report detailed formulas for operating costs and data for "low"- and "high"-cost operating conditions that might be useful in constructing similar formulas for less-typical circumstances. The O & M cost equations do not include any capital or depreciation charges.

The purchase costs predicted by the functions in table 13-2 can be shown to correspond to study-estimate costs for specific designs. The selected designs are quite reasonable for many installations, but the exercise also illustrates the ease with which the estimated costs can diverge in different circumstances. Thus these equations cannot be considered to yield estimates specific to a particular installation, despite the accuracy that may be suggested by the selection of an equation for the actual piece of control equipment.

Edmiston and Bunyard also suggest multipliers to convert the calculated purchase costs into installed costs that include auxiliary equipment and field installation, but not site preparation and other direct and indirect costs. Total-cost multipliers that include all these costs have been estimated by McIlvaine and Ardell (1978). Their factors and the Edmiston and Bunyard factors adjusted to reflect the additional costs are presented in table 13-2. For an average installation the total cost will be obtained by multiplying the calculated purchase cost by the total cost factor.

Flue-Gas-Desulfurization Cost Functions

The importance of flue-gas desulfurization (FGD) for sulfur-dioxide control at electric-power plants has resulted in many studies of control costs and several attempts to characterize these costs in a reasonably accurate yet simple way. Burchard et al. (1972) proposed a cost function recognizing

Table 13-2
Air-Pollution-Control Equipment-Cost Functions
(thousands of dollars)

Equipment Type	Purchase Costs	Total-Cost Multiplier	O & M Costs
Cyclone			
High (95%) efficiency	4.82 + 0.393Q	2.1	0.170Q
Medium (70%) efficiency	3.01 + 0.313Q	2.1	0.170Q
Low (50%) efficiency	0.487 + 0.198Q	2.1	0.170Q
Electrostatic precipitator			
High (99+%) efficiency	84.7 + 1.24Q	2.9	0.133Q
Medium (98%) efficiency	62.4 + 0.881Q	2.9	0.100Q
Fabric filter			
High (400+ °F) temperature	28.9 + 1.67Q	2.2	0.328Q
Medium (250 °F) temperature	6.94 + 0.894Q	2.2	0.328Q
Low (180 °F) temperature	5.31 + 0.650Q	2.2	0.328Q
Scrubber			
High (60 in. H_2O) energy	5.76 + 0.455Q	4.5	2.76Q
Medium (15 in. H_2O) energy	5.76 + 0.455Q	4.5	0.706Q
Low (5 in. H_2O) energy	2.51 + 0.290Q	3.3	0.313Q
Afterburner			
Catalytic (without heat exchanger)	15.1 + 3.03Q	1.8	8.15Q
Direct flame (without heat exchanger)	11.4 + 2.34Q	1.8	15.9Q

Source: Purchase costs and O & M costs adapted from A.S. Kennedy, R.L. Reisenweber, K.G. Croke, and M.A. Snider, *An Economic Comparison of Point-Source Controls and Emission Density Zoning for Air Quality Management,* Air Pollution/Land Use Planning Project, vol. 3, EPA 450/3-74-028c (U.S. Environmental Protection Agency, 1973); Norman G. Edmiston and Francis L. Bunyard, A systematic procedure for determining the cost of controlling particulate emissions from industrial sources, *Journal of the Air Pollution Control Association* 20 (1970): 446-452; and U.S. Environmental Protection Agency, *Control Techniques for Particulate Pollutants* (AP-51). Purchase costs adjusted with M&S index and operating costs adjusted with appropriate components of the U.S. Bureau of Labor Statistics hourly wage and producers' price indexes. Total-cost multipliers from Robert W. McIlvaine and Marilyn Ardell, *Research and Development and Cost Projections for Air Pollution Control Equipment,* EPA 600/7-78-092 (U.S. Environmental Protection Agency, 1978); and adapted from Edmiston and Bunyard, A systematic procedure, based on data in Industrial Gas Cleaning Institute, *Air Pollution Control Technology and Costs in Nine Selected Areas,* APTD 1555 (U.S. Environmental Protection Agency, 1972); idem, *Air Pollution Control Technology and Costs in Seven Selected Areas,* EPA 450/3-73-010 (U.S. Environmental Protection Agency, 1973); and Alan S. Cohen, *An Economic Evaluation of Proposed Amendments to the Illinois Sulfur Dioxide Regulations R74-2, R75-5 and R76-9,* PB 282 390 (Chicago: Illinois Institute for Environmental Quality, 1977).

Note: Units of purchase cost and O & M cost are thousands of 1978 U.S. dollars, and Q is in thousands of actual cubic feet per minute (acfm) of throughput gas. Equations are valid over the range of 10^4-10^6 acfm except for scrubbers, which are valid over the range of (5×10^3)-(2×10^5) acfm, and afterburners, which are valid over the range of (2×10^3)-(2×10^4) acfm. Use of heat exchangers on the afterburners will reduce operating costs by approximately 50 percent. Capital costs for heat exchangers must be calculated separately.

that control equipment often is installed in modules of approximately 4 × 10^5 actual cubic feet per minute (acfm) capacity, which is roughly equivalent to 125 megawatts (MW) of generating capacity. Marder (1977) updated and revised their equations and coefficient sets for four FGD processes to permit calculations of both total equipment costs and annual operating costs. He reports a comparison of costs estimated by his data and equations with estimates from two extensive studies of electric-power-plant FGD systems. The comparison is excellent, with his values falling, in most cases, between the two other estimates.

The Burchard–Marder equation for total FGD capital costs (C_T) is:

$$C_T = \left[36.05 \, D_s F_r \left(\frac{P}{n} \right)^{0.65} n_r^{0.85} + 18.39 \, D_a S^{0.67} \right] I_p (1 + C_I) \quad (13.3)$$

where D_s = The scrubber direct costs in $/kW (dollars per kilowatt).

F_r = A retrofit-difficulty factor.

P = The capacity of the power plant in MW.

n_r = The actual number of modules used.

n = The theoretical number of modules required.

D_a = The scrubbing liquor equipment cost in $/tons S/hr (dollars per ton of sulfur per hour).

S = The design sulfur removal rate in tons S/hr.

I_p = The current CE Plant Index with 1958–1959 = 100.

C_I = The indirect costs as a fraction of direct costs.

In equation 13.3 Marder's redundancy factor has been modified to be consistent with the Burchard et al. definitions of the terms, and the constant terms have been adjusted to remove an "interest-during-construction" factor. The total cost (C_T) will be in current U.S. dollars, since the I_p term will inflate to the chosen evaluation period.

Values for D_s and D_a for the limestone-, lime-, Wellman–Lord, and double-alkali-control processes, with and without additional particulate-matter control and with and without sludge ponding, are given in table 13-3. The theoretical number of modules required is found by dividing the plant capacity by 125 MW and rounding to the nearest integer. This also defines the actual size of the modules used; therefore, attempting to make the design conservative by selecting a higher integer will distort the cost estimate. The need for spare modules for operational redundancy can be accommodated in the n_r term. Often the design will provide one additional module if the theoretical number is one to five, two additional modules if

Data for Estimating Control Costs

Table 13-3
Coefficients for Burchard-Marder Cost Functions

Variable	Limestone	Lime	Wellman-Lord	Double Alkali
D_s ($/kW)				
No particulate control	25.76	25.76	24.23	25.33
With particulate control	35.02	35.02	35.45	36.35
D_a ($/ton S/hr)				
No on-site pond	275.0	250.0	1060.0	325.0
With on-site pond	650.0	580.0	n.a.	700.0
A_s ($/MW)				
No particulate control	3728.0	3728.0	3728.0	3728.0
With particulate control	4728.0	4728.0	4728.0	4728.0
A_a ($/ton S/hr)				
On-site disposal	34.25	64.60	52.90	93.34
Off-site disposal	207.85	196.58	n.a.	225.32
I_a	PPI 0613	PPI 0613	PPI 0543	PPI 0613
I_b	187.2	187.2	235.0	187.2

Source: Adapted from Sidney M. Marder, *Capital and Operating Cost Equations for Flue Gas Desulfurization Devices* (Chicago: Illinois Institute for Environmental Quality, 1977).
Note: PPI 0613 is the U.S. Bureau of Labor Statistics Producers' Price Index for Basic Inorganic Chemicals, and PPI 0543 is the index for Industrial Power.
n.a. = not applicable

the theoretical number is six to ten, and so forth. The retrofit factor will be equal to 1.0 for a new plant and should be set to 1.25 for an average situation at an existing plant.

If the required sulfur removal rate (S) is not known, it can be calculated from

$$S = 0.46 \left(\frac{H_r}{H_c} \right) F_s EP \tag{13.4}$$

where H_r = Heating rate of the boiler in BTU/kWh (British thermal units per kilowatt-hour).

H_c = Heat content of the coal in BTU/lb.

F_s = Fraction of sulfur in the coal.

E = Required control efficiency.

P = Capacity of the boiler in MW.

The equation assumes that 8 percent of the sulfur remains in the bottom ash. If information on the coal and boiler is not available, the (H_r/H_c) term may be assumed to be equal to 0.7 for high-BTU eastern-U.S. coals, 1.0 for most lower-BTU western coals, and 1.2 for lignite coals. The F_s and E terms are expressed as decimal fractions.

The "with-particulate-control" coefficients in table 13-3 are based on a venturi scrubber as the particulate-control device. If it is desired to estimate costs with different equipment for particulate control, the FGD control should be estimated from equation 13.3 without particulate control and the appropriate cost for a fabric filter or electrostatic precipitator from table 13-2 added on. The approximate conversion of 2 standard cubic feet per minute of exhaust gas per kilowatt of capacity may be used.[2] The costs for on-site disposal assume the material will be clarified and that the pond is lined. The Wellman–Lord process includes the cost of facilities for production of sulfur.

The Burchard–Marder equation for operating costs (C_{OM}) is:

$$C_{OM} = \left[A_s P \left(\frac{I_s}{235} \right) + 8{,}760\, A_a S \left(\frac{I_a}{I_b} \right) \right] L + C_L + MLC_T \quad (13.5)$$

where A_s = Annual scrubber utilities cost in \$/MW.

P = Capacity of the power plant in MW.

I_s = A price-adjustment factor.

A_a = Scrubbing-liquor handling and raw-materials cost in \$/ton S.

S = The design sulfur-removal rate in tons S/hr.

I_a, I_b = Price-adjustment factors.

L = The yearly load factor.

C_L = An annual labor and overhead cost.

M = A fractional maintenance factor.

C_T = The total capital cost.

The price-adjustment factors are the U.S Bureau of Labor Statistics Producers' Price Index for Industrial Power (no. 0543, 1967 = 100) and for Basic Inorganic Chemicals (no. 0613, 1973 = 100). No. 0543 is used for I_s for all processes. The assignment for I_a is shown in table 13-3. Values for A_s and A_a also are given in table 13-3.

The load factor will decrease as the plant ages and becomes less efficient. A detailed schedule is provided by McGlammery et al. (1975). As an approximation, 0.75 may be used for new plants and 0.55 for existing, older plants. The annual labor and overhead cost may be taken as \$250,000 (in 1977 U.S. dollars) for all systems. The fractional maintenance factor may

Data for Estimating Control Costs

be set to 0.11, and will include scheduled replacement and insurance as well.

The values given for A_a for off-site disposal assume fixation of the sludge and movement by truck. The A_a value for the Wellman–Lord process includes a sulfur-byproduct credit of $40 per ton.

Making a Study Estimate

The greater accuracy of a study estimate is simply a result of the greater specificity of the information developed in the analysis. Each source of pollutant emissions must be identified, a required level of control defined, a schematic design of the necessary control system laid out, and calculations made for each item of equipment in the control program. If the control equipment will be installed at an existing plant, it will be prudent to visit the site; to complete a rough sketch of the equipment locations to define the length of ducts and ensure that site preparation will not require demolition and reconstruction of existing facilities; and to determine whether any unusual aspects of the facility, the exhaust gases, or the raw materials will affect the design of the control equipment.

The greatest difficulty in making a study estimate is in collecting cost data for the various pieces of equipment and other required operations, such as site development, without spending as much time as would be required to solicit quotes directly from suppliers. A few comprehensive sources of data are available. One of the more useful is *Capital and Operating Costs of Selected Air Pollution Control Systems* (Neveril 1978).[3] This report contains cost functions for most types of pollution-control equipment and for a number of other major pieces of equipment that are commonly a part of an air-pollution control program. It also provides data on typical values of the variables needed to make the calculations, clear explanations of the use of the cost functions, data for operating costs, and worked-out examples. Another comprehensive source of cost data is the lengthy series of articles entitled "Evaluation of Capital Cost Data" appearing in the *Canadian Journal of Chemical Engineering*. This series has presented cost data on motors, fans, pumps, utilities, and other auxiliary equipment (for example, Woods, Anderson, and Norman 1976, 1978), with several articles on air-pollution-control equipment forthcoming. These articles contain data for many types of equipment that often are difficult to find in the literature. The data are presented both graphically and as base cost and exponent for use with equation 13.1.

Cost data for specific control devices and auxiliary equipment, or for equipment to be used under difficult or unusual circumstances, will require a careful literature search. As recommended in chapter 12, the *Control Techniques* series and the support documents for the New Source Performance Standards published by the U.S. Environmental Protection Agency often yield useful references to the cost literature. Another useful source of data and references is the eight-volume *Technology Assessment Report for Industrial Boiler Applications* (U.S. Environmental Protection Agency 1979d), which describes and provides cost data not only for controls for sulfur oxides, particulate matter, and nitrogen oxides, but also for alternative combustion technology and fuel-cleaning and -switching techniques. Much of the information cited there is applicable to many other pollutant sources.

Several additional sources of cost information on control devices and auxiliary equipment can be recommended. Detailed data for estimates of spray, tray, and packed towers are contained in Peters and Timmerhaus (1979), Pikulik and Diaz (1977), Blecker and Nichols (1973), and Guthrie (1970). Useful data on both capital and operating costs of thermal and catalytic incinerators have been published by C-E Air Preheater (1976). Cost functions for many items of auxiliary equipment, particularly solids and liquids handling equipment, are presented by Peters and Timmerhaus and by Blecker and Nichols. Finally, comprehensive cost data for sludge handling and disposal are available in Michael Baker (1980).

An alternative to developing a schematic design and searching out cost functions for the diverse equipment required is available for FGD and particulate control for coal-fired electric-power plants. Two reports have assembled all the information needed to complete the calculations. Ponder et al. (1976) provide the information required for estimates of the total capital and operating costs of lime, limestone, double-alkali, magnesium-oxide, and Wellman–Lord sulfur-dioxide scrubbing systems. Information also is provided to allow the inclusion of a venturi scrubber in the system for additional particulate control. This report is arranged in a workbook format with nomographs and line-by-line instructions that make the calculations simple to follow. An abbreviated calculation procedure is included to permit a quick estimate of at least order-of-magnitude accuracy. A report by the U.S. Environmental Protection Agency (1980d) is similarly complete, but approaches the calculations by reporting the cost of complete modules. The total cost is then determined by the number of modules (plus spares) needed to control the gas flow. This report provides data for calculation of costs for nitrogen-dioxide control; particulate-matter control by electrostatic precipitators, fabric filters, and venturi scrubbers; and sulfur-dioxide control by lime and limestone wet scrubbers and by semidry scrubbers. Both reports contain information on auxiliary equipment and operating costs that would be useful for calculations in other situations.

Computer programs are available that execute both the control-equipment design and cost analysis for air-pollution control on coal-fired electric-power plants. The Shawnee Lime-Limestone program developed by the U.S. Tennessee Valley Authority estimates the capital and operating

Data for Estimating Control Costs

costs of a complete FGD system, including sludge disposal. Either a venturi or a turbulent-contact-absorber scrubber with either lime or limestone scrubbing liquors may be selected for analysis. The program requires the efficiency of the particulate-control device upstream of the FGD system to be specified and does not estimate the costs of particulate control. It also requires a number of important design variables for the scrubber as well as the boiler and coal characteristics to be supplied by the user. A complete description of the program and a list of the necessary data is given by Torstrick, Henson, and Tomlinson (1978).

The program developed by PEDCo Environmental (1978) includes particulate-control costs for an electrostatic precipitator, fabric filter, or venturi scrubber; and turbulent-contact-absorber scrubbing with lime, limestone, double-alkali, Wellman–Lord, and magnesium-oxide systems. This program also requires a number of the design variables for the particulate-control and sulfur-dioxide-control systems to be supplied by the user.

A cost-analysis program being developed by the Argonne National Laboratory combines a particulate-cost-estimation program and an FGD program that contains all the options of the previously described programs, two additional FGD control systems, and a number of sludge-disposal and byproduct options (Farber 1977 and Farber and Livengood 1979). A coal-characteristics data base is associated with this program so that only the geographic source of the coal is required. A few design variables for the control equipment must be supplied.

The Particulate Control Performance and Cost Model (*PC2M*) developed by Teknekron (Chapman et al. 1980) is intended for cost-minimizing design studies of control of particulate-matter emissions from plants firing low-sulfur, western coals. The only data needed are the boiler, coal, and ash characteristics and the emission requirements that must be met. The program makes all the other necessary calculations.

Each of these programs (and often many cost functions reported in the literature) report annual operating and maintenance costs with a portion of the capital costs included as depreciation charges. It is not clear from their documentation whether it is possible to obtain only the operating and maintenance costs; but because the factors used for calculating the annualized capital costs are stated, the capital costs could be manually subtracted from the reported results.

Other Cost Data

Values for many of the items of direct and indirect costs listed in table 12-1 can be very difficult to estimate with any accuracy. In making an order-of-magnitude estimate, the problem is avoided by using a total-cost multiplying factor, such as those presented in table 13-2. A study estimate will require some of these items to be estimated directly, although others will be entered as a percentage of the equipment-purchase costs or the total of the direct costs. Even the items based on a percentage of other costs should be examined individually and adjusted to reflect the specific circumstances of the project being evaluated.[4]

Site-preparation costs will have to be directly estimated for an existing plant. Guthrie (1969) suggests costs for most of the operations that might be required. For a new facility, site preparation can be assumed to be negligible. In either case, land costs also must be estimated. Even if the land is presently unused, occupying the site with the control equipment forecloses other uses in the future.

Working capital is required to support the operation of the control equipment. At a minimum it will include the value of the raw materials and by-products kept on hand. Several analysts have suggested using three to four weeks' supply of raw materials and byproducts and seven weeks' supply of wages and overhead as an estimate.[5]

The value of the spare-parts inventory has been taken by several analysts to be 0.5 to 1 percent of the value of the purchased equipment, although 2 percent would be reasonable for an FGD system.

The costs of an ambient monitoring network also are specific to each project. Monitoring may be conducted by the enforcement agency without any change in its existing network, or an entirely new network may be necessary. Table 13-4 gives the costs of installing and operating monitoring stations and a monitoring network. This table also gives the costs of establishing a meteorologically determined intermittent (or supplementary) control system (ICS), which can be used to reduce the likelihood of local high air-pollution concentrations by curtailing emissions during adverse weather conditions. Additional data on the cost of an ICS are given by Leavitt et al. (1974).

The additional enforcement costs of the agency will include engineering and staff support to evaluate any permit applications that may be required by the proposed program, to maintain surveillance of the sources, and to conduct annual inspections of the sources. A computer program, the Air Pollution Control Strategy Resource Estimator, has been developed by the Argonne National Laboratory to estimate the staff and other requirements that would be demanded by the enactment of a new regulation (Senew et al. 1978). Either the model, or factors drawn from it, could be used to estimate enforcement costs for other situations.

Most of the remaining costs listed in table 12-1 can be estimated from factors given by Neveril (1978). He offers reasonable average percentage

Table 13-4
Air-Pollution-Monitoring Network and Intermittent-Control-System Costs
(thousands of dollars)

	Network Fixed Costs	Variable Costs Per Site
Initial costs		
System design[a]	$125	
Site selection[a]	25	
Site acquisition	6	1
Training/startup[a]	63	
Shelters		4
Monitors and auxiliaries		
TSP hi-vol		0.5
Sulfur dioxide		
Bubbler		1
Continuous		9
Carbon Monoxide		7
Ozone		8
Nitrogen dioxide		
Bubbler		1
Continuous		11
Spares and spare parts		25% of Monitors and auxiliaries cost
Meteorology tower[b]	61	
Telemeter systems	51	
Operating (annual) costs		
ICS consultants[a]	240	
Operating and maintenance	9	3
TSP hi-vol		1
Sulfur dioxide		
Bubbler		3
Continuous		7
Carbon monoxide		7
Ozone		7
Nitrogen dioxide		
Bubbler		2
Continuous		8
Telemeter lines		1
Supervision	40	

Source: Adapted from A. K. Miedema, C.E. Decker, F. Smith, and J. White, *Cost of Monitoring Air Quality in the United States*, EPA 450/3-74-029 (U.S. Environmental Protection Agency, 1973); and Wayne E. Wesolowski, *Initial Development and Operating Costs for Supplementary Control Systems* (Chicago: Illinois Institute for Environmental Quality, 1977).

Notes: In thousands of 1978 U.S. dollars. Total variable cost per site is the sum of site and shelter costs and costs of individual pollutant monitors located at that site. Total network cost is sum of network fixed costs and total variable cost per site for all sites.

[a]Costs shown are for an intermittent-control system (ICS). Monitoring networks not associated with an ICS will be substantially less.

[b]Cost is for one tower.

values specific to all the common tail-end control devices, and presents a table of factors that enable the analyst to adjust the individual cost items to reflect better the circumstances of the project being evaluated. The only items that appear to be low are "startup," which is set at 1 to 2 percent but has been estimated by other analysts as high as 10 percent, and the cost of piping for a venturi scrubber, which is set at 5 percent but has been reported by others as between 30 and 40 percent of purchased-equipment costs.

Estimating Transportation-Pollutants-Control Costs

Emissions from motor vehicles may be reduced by controlling the amount of pollutants released by each vehicle per mile of travel, by reducing the number of vehicles in use, or by reducing the amount of travel by each vehicle. Vehicle emissions are controlled by engine modifications or by treating the exhaust gas. As the regulations have become more stringent, the necessary control equipment has become more complex and more expensive. Dewees (1974) gives a history and cost analysis of the development of automobile-emission-control systems in the United States. Estimates of the initial purchase cost and annual operating costs also are reported by Schwing et al. (1980). A detailed analysis of the capital costs is provided by Lindgren (1978).[6]

In estimating the total cost for a vehicle population, it is necessary to consider the amount the average car of each age is driven as well as the fraction of the vehicles of a particular model year that are in use. National average data are reported by the U.S. Environmental Protection Agency (1978e), although there is substantial variation among local areas. It will be necessary to establish initial values for the control equipment already in place in the vehicle population at the beginning of the analysis, as well as terminal values at the end of the analysis period.

Since regular maintenance is required to keep the exhaust-control equipment working effectively, many local governments have adopted programs of periodic vehicle inspection. Detailed information on the costs of inspection and maintenance programs is reported by Kincannon and Castaline (1978) and Palmini and Rossi (1980).

A wide variety of projects designed to reduce the amount of traffic have been attempted in cities throughout the world. They range from pedestrian-only areas to car-pool ride-matching services. Some costs have been reported for transportation-control projects; but the great differences in project design, even for projects that are conceptually similar, limit the usefulness of available data. Projects that involve construction within an existing road right-of-way are particularly prone to unusual, site-specific costs. Useful data for order-of-magnitude estimates have been published by the U.S. Environmental Protection Agency (1978f) for car-pool programs and for separate transit lanes and freeway access ramps. Keyani and Putnam (1977) report cost experience for these projects and for several other programs, such as parking-rate controls, bikeways, fringe parking lots, traffic-signal synchronizaion, and shelters for transit riders.

Notes

1. The value of 0.6 is based on the physical relationship between the surface and volume of spherical and cylindrical structures (Williams 1947). However, it is only a rough approximation, which should be used only if no other information is available.

2. The equations in table 13-2 require actual cfm rather than standard cfm. An electrostatic precipitator operated ahead of the power-plant heat exchanger can be estimated to be at 650 °F and an electrostatic precipitator or fabric filter operated downstream of the heat exchanger can be taken as at 300 °F. If standard conditions are 70 °F, then the factor is (T °F + 460)/530 to convert scfm to acfm.

3. The Neveril (1978) report is a revision and extension of the better-known series of articles by Neveril, Price, and Engdahl (1978). The cost functions for incinerators and carbon adsorbers have been substantially revised. Cost functions for gas-absorption towers, refrigeration, flares, and some auxiliary equipment, as well as considerable explanatory material, have been added.

4. A useful source of estimates for these costs is Devitt, Spaite, and Gibbs (1979).

5. For further discussion of working capital see Weaver and Bauman (1973) and Scott (1978).

6. Costs for Canadian vehicles can be estimated from the data presented by these studies if 1971 U.S. costs are used for the 1971-1974 Canadian model years and 1973 U.S. costs for 1975-1980 model years.

Appendix A: Sources, Measurement, and Effects of Air Pollutants

This appendix summarizes information on most of the air pollutants that are now or are expected soon to be regulated by the U.S. or Canadian federal authorities. It describes the physical properties of the pollutants, the primary sources of the pollutant in the ambient air and the common means of their control, the prevalence of the pollutant and some information on how those concentrations have been measured, and a summary of the various health and environmental effects.

A recent review that contains additional information on the pollutant is listed at the end of each entry. More-detailed information is available from several sources. The U.S. Environmental Protection Agency (EPA) is now preparing five revised *Air Quality Criteria* documents for the six pollutants regulated by the ambient-air standards of the U.S. Clean Air Act, beginning with lead in 1977 and concluding with sulfur dioxide and particulate matter in 1981. Detailed reports on pollutants regulated as "hazardous pollutants", such as vinyl chloride, have been published by the EPA under several titles. In general, these documents are the most comprehensive source of information on the regulated pollutants. The National Research Council of the U.S. National Academy of Sciences has released twenty-six studies of air pollutants. These studies often cover pollutants that are not now regulated. The title of each volume is the name of the pollutant. The U.S. National Institute for Occupational Safety and Health has issued some ninety documents on air contaminants entitled *Criteria for a Recommended Standard: Occupational Exposure to* (followed by the name of the pollutant). These reviews cover only health effects and include only limited information on the effects that might be observed at the lower concentrations characteristic of community exposures. More-concise reviews of the effects of air pollutants have been published from time to time as journal articles (for example, Ferris 1978), or even as monographs from special-interest organizations, such as the one issued by the American Lung Association (Shy et al. 1978). Comprehensive handbooks on air pollution often include an article or series of articles on the effects of various pollutants (for example, Stern 1976, 1977a).

Arsenic

Physical Form. Arsenic is found as a fine particle or condensed on the surface of other particles at normal ambient temperatures. Arsenic trioxide is of most concern, although there is also concern about forms such as lead arsenate.

Major Sources. Arsenic air pollution originates primarily from copper, lead, and zinc smelters processing ores containing arsenic and from the manufacture and spraying of certain insecticides and herbicides. Arsenic emissions from nonferrous smelters can be controlled by cooling the exhaust gas so that the arsenic vapors will condense into particles that can then be filtered out in a baghouse.

Ambient Concentrations. Significant community exposures to arsenic are limited to the areas adjacent to a few smelters. Measurements of arsenic are made by collecting the particles on a membrane filter and analyzing by atomic-absorption spectroscopy.

Health Effects. Arsenic compounds are believed to be a cause of lung cancer. A significantly increased incidence of lung cancer has been reported among adults living near a copper smelter.

Further Information. National Research Council (1977).

Asbestos

Physical Form. Asbestos occurs as a filimentous particle, potentially no longer than 1 micrometer.

Major Sources. Asbestos is widely used as insulation on various building materials, especially in buildings built between 1930 and 1972. It can be released into the ambient air either by erosion from surfaces or during demolition activities. Emissions from demolition can be reduced by observing certain work practices. The most-concentrated sources are industrial plants that manufacture or use asbestos fibers. Emissions are controlled by filtering the exhaust gases through a baghouse.

Ambient Concentrations. The mining and milling of asbestos fiber is a very small industry. Asbestos is measured by collecting the fibers on a membrane filter and counting them under an optical or electron microscope.

Health Effects. Exposure to asbestos particles can cause extensive scarring of lung tissues (asbestosis), which may be associated with breathlessness and productive cough. Even when asbestosis does not develop, lung cancer can be caused by airborne asbestos. Cigarette smoking increases the risk of lung cancer from asbestos. Apparently cancer can be induced at very low exposure levels.

Further Information. Zielhuis (1977).

Benzene

Physical Form. Benzene is a colorless, highly volatile liquid that will usually exist as a gas in the ambient air.

Major Sources. The principal source of benzene in the ambient air is gasoline, specifically the evaporation from gasoline in automobiles, gasoline storage and marketing facilities, and petroleum refineries. Benzene is important as an intermediate in the manufacture of industrial chemicals, and emissions from such facilities are a major source. Emissions may be controlled by adsorption on activated carbon or by incineration.

Ambient Concentrations. Benzene is found in the air near heavily traveled streets, service stations, and some chemical manufacturing plants, especially those producing malic anhydride and aniline. Benzene is measured by gas chromatography.

Health Effects. The primary health effects of benzene exposure appear in the blood-forming systems. Prolonged exposure results in a substantial reduction in the number of red blood cells. In some cases, death has resulted from leukemia. Injury to chromosomes also has been reported.

Further Information. U.S. National Institute for Occupational Safety and Health (1974).

Cadmium

Physical Form. Cadmium is usually found as a fine particle or condensed on the surface of other particles at ambient temperatures.

Major Sources. Cadmium air pollution originates primarily from the incineration of plated metals and from ferrous and nonferrous smelters processing ores and scrap containing cadmium. Cadmium emissions can be controlled by filtering exhaust gases through a baghouse.

Ambient Concentrations. Significant community exposures to cadmium are limited to the areas adjacent to a few smelters. Cadmium is measured by collecting the particles on a membrane filter and analyzing them by atomic absorption spectroscopy.

Health Effects. Exposure to cadmium over a period of time can result in irrevesible damage to the kidneys. In some community exposures, softening and spontaneous fractures of the bones have been observed. Cadmium has been shown to cause cancer in animals.

Further Information. U.S. National Institute for Occupational Safety and Health (1976).

Carbon Monoxide

Physical Form. Carbon monoxide is a colorless, odorless, tasteless gas.

Major Sources. Indoor sources of carbon monoxide are gas and wood cooking and heating stoves and cigarette smoking. In most urban areas, 85 percent or more of the outdoor carbon-monoxide pollution comes from motor vehicles. Incomplete combustion in industrial processes and boilers and in home wood stoves also is an important outdoor source of carbon monoxide. Industrial emissions often are controlled by burning the carbon monoxide to recover the waste heat. Motor-vehicle emissions are controlled by small catalytic incineration units.

Ambient Concentrations. Elevated concentrations of carbon monoxide are found almost exclusively near busy or congested traffic. Concentrations fall off very rapidly with distance from the traffic lanes, by as much as a factor of four at 200 feet. Carbon monoxide is measured by continuous-gas-phase nondispersive infrared spectrometry.

Health Effects. Carbon monoxide is absorbed into the blood through the lungs, displacing oxygen in the red blood cells, forming "carboxyhemoglobin," and reducing the amount of oxygen available to the cells throughout the body. When 4–5 percent of the blood hemoglobin is carboxyhemoglobin, mental processes are adversely affected and the ability to perform work is decreased in normal adults. Decreased tolerance for exercise has been observed at 1.5–3 percent blood carboxyhemoglobin in persons with heart disease, chronic obstructive pulmonary disease, and certain diseases of the blood vessels.

An adult woman at rest at sea level will build up a carboxyhemoglobin level of about 2 percent when exposed to 35 ppm of carbon monoxide for

Appendix A

one hour or when exposed to 12 ppm for eight hours. If she were exercising when exposed to 35 ppm, her blood carboxyhemoglobin would reach about 3 percent in an hour.

Further Information. U.S. Environmental Protection Agency (1979a).

Fluoride

Physical Form. Fluoride may occur as a gas as fluorine or hydrofluoric acid (HF), or as a particle as a compound such as calcium fluoride.

Major Sources. The primary sources of fluoride in the ambient air are aluminum-reduction facilities and phosphate fertilizer, phosphoric acid, and phosphorous-manufacturing plants.

Emissions from aluminum-reduction facilities have been controlled by mixing dry aluminia with the gas, which absorbs the fluoride gas, and then removing the particles in a baghouse. This process usually results in raw-materials savings substantially in excess of the cost of installing and operating the control equipment. Emissions from phosphate fertilizer and phosphorous plants are generally controlled by wet scrubbers.

Ambient Concentrations. Community exposures to fluoride generally are limited to areas adjacent to the sources listed previously and a relatively few other industrial sources. Fluoride particles and gas are measured by drawing the air sample through a chemically pretreated paper and analyzing the paper by wet-chemistry spectrophotometry.

Health Effects. There is no evidence of adverse health effects due to fluoride air pollution in North America, although effects have been observed in occupationally exposed workers. Skin sores and reduced hemoglobin levels have been reported in Europe among residents near industrial sources with no occupational exposure.

Environmental Effects. Fluoride is the most phytotoxic of all the common air pollutants. Injury to vegetation usually is associated with gaseous fluoride, which is absorbed by the leaves. Although airborne fluoride particles are not absorbed in sufficient quantity to injure plants, they may accumulate on leaves and other surfaces. Adverse effects may occur in foraging animals that consume the contaminated vegetation.

Unlike other common pollutants, fluoride accumulates in plants, causing tip and edge burn in the leaves or needles and reducing the yield or the rate of growth. Injury to needles has been observed in conifers from exposures of 1.2 ppb (parts per billion) for one to two days. Reduced yield was observed in broad-leaf crops at 0.6 ppb average exposure over a growing season.

Reduced milk production and lower growth rates have been observed in cattle fed on forage containing fluoride at 50 $\mu g/g$. Lowered growth rates have been observed in cattle feeding over a lifetime on forage with less than 30 $\mu g/g$. Accumulation of fluoride of about 120 $\mu g/g$ in honey bees was found to be toxic.

Further Information. Hughes (1977).

Hydrogen Sulfide

Physical Form. Hydrogen sulfide is a colorless gas with a rotten-egg smell.

Major Sources. Hydrogen sulfide is released from several industrial processes, particularly pulp and paper mills, oil refineries, oil- and gas-production fields, and municipal waste-water-treatment ponds. The natural decay of plant material is also a major source. Exhaust gases containing high concentrations of hydrogen sulfide can be controlled by a Claus-process plant that will produce pure sulfur from the hydrogen sulfide. Lower-concentration exhaust gases can be controlled by incinerating the gas in a boiler (often with a significant heat recovery). The resulting sulfur dioxide is then controlled by a wet absorber.

Ambient Concentrations. Hydrogen-sulfide odors may permeate through an entire town where there is a strong industrial source. Hydrogen sulfide is measured by drawing the air sample through a chemical solution and analyzing by wet-chemistry spectrophotometry.

Health Effects. The odor of hydrogen sulfide can be detected at about 4.5 ppb. Very little documentation exists of community reactions to specific concentrations of hydrogen sulfide. Numerous complaints of nausea, headache, and shortness of breath were recorded in one study when hydrogen-sulfide concentrations were approximately 300 ppb.

Environmental Effects. Both visible damage and growth and yield loss have been observed in timber, sugar beets, and alfalfa at concentrations of 100–300 ppb of hydrogen sulfide for four weeks to eight months. Increased sensitivity to the pollutants was observed when bean plants were exposed to both ozone and hydrogen sulfide. At similar exposures, damage to some house paints is observed.

Appendix A

Further Information. National Research Council (1978).

Lead

Physical Form. Lead is found as a fine particle or condensed on the surface of other particles.

Major Sources. Except near lead smelters and other nonferrous mining and smelting operations, almost all ambient lead is from motor vehicles burning gasoline containing lead additives. The EPA has established regulations to reduce the amount of lead in gasoline in the U.S. Lead emissions from nonferrous smelters can be controlled by a baghouse.

Ambient Concentrations. Lead air pollution occurs exclusively in urban areas near heavy traffic and in the vicinity of a few smelters. Lead is measured by collecting the particles on a glass-fiber filter and analyzing by atomic-absorption spectroscopy.

Health Effects. Inhaled lead particles are deposited either in the throat or in the lungs. In children, as much as 50 percent of the inhaled lead is eventually absorbed into the blood. Lead can also be taken in through food and water and, in the case of children, through soil. Indications of a disruption in normal hemoglobin production are observed in children when blood lead concentrations exceed 200 $\mu g/L$ (micrograms per liter). The onset of anemia is observed at approximately 400 $\mu g/L$. Significantly lower mental ability among children with blood lead concentrations above 350–400 $\mu g/L$ has been reported.

In an area with moderate concentrations of lead in the soil approximately 7 percent of the children were found to have blood lead in excess of 400 $\mu g/L$ when air lead concentrations were above 1.5 $\mu g/m^3$.

Environmental Effects. Because horses tend to eat roots and soil as they graze, lead poisoning of horses can be a problem in areas with soil contaminated by lead air pollution.

Further Information. U.S. Environmental Protection Agency (1977a).

Nitrogen Oxides

Physical Form. Nitric oxide is an odorless, colorless gas. Nitrogen dioxide is a yellow-brown gas with a sweet, pungent odor. Both can be oxidized in the air to nitrates, which may be present in either gaseous or particulate form. Although technically incorrect, the term *nitrate* is used to include the nitrous-acid and nitrite forms. Nitrogen oxides also can react with simple amine molecules to form nitrosamines.

Major Sources. Nitric oxide and nitrogen dioxide are produced by motor vehicles and stationary combustion sources, such as coal- or oil-fired power plants and industrial boilers. Nitrogen oxides are created by a reaction between atmospheric nitrogen or nitrogen in the fuel and atmospheric oxygen in the furnace or combustion chamber. Emissions of nitrogen oxides are reduced by controlling temperatures in the combustion process to limit their formation. Natural sources of nitrogen oxides may exceed half of the total emissions. Gas cooking stoves are a major indoor source of nitrogen dioxide.

Ambient Concentrations. Elevated concentrations of nitrogen dioxide are found almost exclusively in or near urban areas or near stationary sources. The less-harmful nitric oxide is rapidly converted to nitrogen dioxide where high concentrations of ozone from photochemical smog are present. Otherwise, the reaction proceeds slowly.

Prior to 1973 most measurements of nitrogen dioxide were made by the Jacobs–Hochheiser method, which has been found to produce significant errors, frequently overstating the concentration of nitrogen dioxide in the air. Nitrogen dioxide is now meaured by continuous-gas-phase chemiluminescence. Nitrates are measured by collecting particles on a glass-fiber filter and analyzing by wet-chemistry spectrophotometry.

Health Effects. Inhaled nitrogen dioxide interferes with the defense mechanisms of the lung, increasing susceptibility to lung infections. Laboratory animals exposed intermittently to 0.5 ppm nitrogen dioxide for 90 days suffered increased mortality when exposed to a pneumonia infection. Studies of children living near a large stationary source of nitrogen dioxide found significantly higher rates of bronchitis and other respiratory diseases in neighborhoods with annual average nitrogen-dioxide levels above 0.08 ppm. Children exposed to frequent indoor levels of approximately 0.2 ppm, one hour average (from gas cooking stoves) were observed to experience increased respiratory disease.

Increased aggravation of heart- and lung-disease symptoms have been observed among elderly patients when nitrate concentrations were greater than 4 $\mu g/m^3$. However, it is not clear whether nitrates are the principal agent or only coincident with other fine particles present at the same time. Nitrates also interfere with the oxygen-carrying capacity of the blood.

Appendix A

Inhalation of nitrosamines has been shown to cause cancer in animals and is presumed to be carcinogenic in humans.

Environmental Effects. Yield reduction is observed in vegetables from chronic exposure to nitrogen dioxide at 0.25 ppm. Reduced yields are observed in orange trees, apparently the most sensitive crop, from chronic exposures at 0.06 ppm. Nitrogen oxides are an important source of increased acidity in rainfall, which may result in increased acidity of lakes and streams. Discoloration of plumes and haze layers due to high concentrations of nitrogen dioxide can degrade visibility. Nitrogen dioxide is known to cause fading in some cloth dyes.

Further Information. U.S. Environmental Protection Agency (1980a).

Odors

Physical Form. All odors are gases.

Major Sources. There are innumerable different sources of offensive odors, but some of the most common are: (1) refuse, organic waste, and rendering; (2) petroleum refining; and (3) pulp and paper manufacture. Some of the most objectionable odors are from hydrogen sulfide (rotten egg), mercaptans (skunk), trimethylamine (fishy), and organic sulfides (skunk). The most effective control for odors is modification of the process or work procedure that produces the odor. Odors are often controlled by thermal or catalytic incineration, by wet absorption, or by adsorption. Inadequate oxidation may increase the problem.

Ambient Concentrations. Where there is a strong source, some odors will pervade an entire town. Sufficient dilution of the odor with surrounding air will reduce the concentration below a detection threshold. Odors are measured by mixing the odorous air and clean air in a breathing mask to determine the amount of dilution necessary to reduce the odor below the detection level.

Health Effects. At low concentrations an odor may be only annoying, whereas at higher concentrations it may cause nausea, headaches, or other adverse effects. The specific effects will vary with the specific odorant. One study suggested that odors more concentrated than seven times the detection threshold would cause complaints to the authorities and another study found that at about eight times the threshold half the members of the community said they were annoyed by an ambient odor.

Further Information. Schroeder (1975).

Ozone

Physical Form. A colorless gas with a sharp smell.

Major Sources. Ozone and other oxidants are produced by photochemical reactions in the atmosphere involving nitrogen oxides and hydrocarbons. The resulting mixture of pollutants generally is called *photochemical smog*. Motor vehicles, petroleum refining and marketing, and organic-solvent use are the major sources of hydrocarbons. Vegetation also may be a significant source. Hydrocarbon emissions can be controlled by thermal or catalytic incineration (often with significant heat recovery) or by adsorption (often with significant raw-materials or product recovery). In some areas photochemical reactions may be limited by the amount of nitrogen oxides present rather than that of hydrocarbons. Ozone occurs naturally in the stratosphere and may be brought down to lower elevations during certain weather conditions, particularly in the late winter and early spring in northern latitudes.

Ambient Concentrations. Because the photochemical reaction requires time to occur, the highest concentrations of ozone and other oxidants will be found downwind of urban areas. In some parts of the United States, large air masses of high-ozone air have been observed to move around with the winds over a region including several states.

Ozone is measured by continuous-gas-phase chemiluminescence. Prior to 1970 a wet-chemistry technique that measured total oxidants, not just ozone, was used in some cities. There is no consistent relationship between the amount of ozone and the amount of total oxidants present in the atmosphere of different cities or even that of a single city; therefore, measurements by different methods should be closely examined before they are used together.

Health Effects. Ozone directly attacks cells, paralyzes cilia, increases secretion of mucus, and can "shut down" portions of the lung at moderate concentrations. Resistance to infection is reduced significantly in laboratory animals at concentrations of 0.10 ppm. Short-term exposures to ozone at 0.26–0.5 ppm have been shown to injure lung tissue and to reduce oxygen-exchange capacity in laboratory animals. Healthy young men engaged in intermittent exercise have been observed to suffer a decrease in lung func-

Appendix A

tion during exposure to 0.37 ppm of ozone for two hours. Chest discomfort was observed in sensitive persons at 0.15 ppm of ozone when exercising. Natural repair mechanisms exist for ozone damage. Partial adaptation to moderate ozone levels has been observed in residents of the Los Angeles, California basin. Eye irritation is almost always associated with photochemical smog, but the cause is more likely to be formaldehyde, acrolein, or peroxyacetyl nitrate (PAN) than ozone.

Environmental Effects. Yield reductions in root crops have been recorded from short-term exposures to ozone concentrations of 0.25 ppm. Chronic exposures of 0.05 ppm have been observed to cause a yield loss of 50 percent in alfalfa, a 10-percent yield loss in corn, and a 25–50-percent yield loss in potatoes. Long-term exposures at 0.10 ppm have resulted in a 10–20 percent growth loss in conifers. Aspen, ash, and azalea are among the more-sensitive woody plants. Approximately 30 percent of the Ponderosa Jeffry pine forests immediately downwind of Los Angeles have been severely damaged by photochemical smog. Ozone exposure at 0.05 ppm causes rubber compounds to become brittle.

Further Information. U.S. Environmental Protection Agency (1978a).

Particulate Matter

Physical Form. Particulate matter may be liquid (droplets) or dry particles from 0.01 μm (micrometers) to 100 μm in diameter (a human hair is about 100 μm in thickness). The greatest number of particles are the "fine" particles (less than 3 μm), whereas the most mass is in the 10–20- μm size range. Fine particles may remain suspended in the air for a very long time, but particles larger than 25 μm rapidly fall to the ground.

Major Sources. Particles larger than a few micrometers are most often produced through mechanical wearing or grinding operations (for example, cement dust, sawdust, and road dust). Where coal-fired boilers lack adequate air-pollution controls, some residues of combustion (that is, fly ash) also may be in this size range. Large particles may be controlled by cyclones. Fine particles (less than 3 μm) are usually the products of combustion; condensation from a gas (for example, iron-foundry smoke); or chemical reactions in the atmosphere (for example, sulfates, nitrates, and particulate organics). Emissions of fine particles can be controlled by a baghouse, an electrostatic precipitator, or a high-energy wet scrubber. The mineral-products, iron and steel, and nonferrous-metals industries are major sources of particulate matter. Stationary fuel combustion (including home heating) is equally important as a source of particulate matter, but a larger percentage of the particles are in the smaller size ranges. Motor vehicles also produce a significant mass of small particles from combustion exhaust and large particles from roads. Open fires are another major source of small particles. The major sources of sulfur oxides, nitrogen oxides, and polycyclic organic matter (gases that form particles in the atmosphere) are discussed under those entries. Important natural sources of particles are sea spray, soil dust, and plants (pollen and spores). The principal indoor source of particulate pollution is smoking.

Ambient Concentrations. Because large particles settle quickly, they are primarily observed close to urbanized areas and human activity. Fine particles can be transported great distances. Natural background concentrations of particles may be 1–30 $\mu g/m^3$, depending on the location and climate. Much of the eastern United States and most of Western Europe is blanketed by a fine-particle haze. As much as two-thirds of the fine particulate matter may be sulfates in some areas. Condensed organic compounds are an important fraction of the fine particles. Fine carbon particles from various combustion sources also have been identified as a major constituent of urban fine particulate matter. Lead and other metals can be a significant component of urban particulate matter. These compounds and others common to transportation and industrial activity tend to be found among the fine particles, whereas the elements common to the earth's crust (such as iron, aluminum, calcium, and silicon) tend to be found among the large particles.

In Canada and the United States, particulate matter is measured by the "hi-volume" method, which essentially consists of a vacuum-cleaner motor behind a glass-fiber filter. This device collects all the particles in the air drawn into the instrument and through the filter. The concentration of suspended particulate matter then is determined by weighing the deposit on the filter and dividing by the volume of air drawn through the filter. Particulate matter measured in this way is called *total suspended particulate* (TSP). Because a few large particles will outweigh millions of fine particles, this method gradually is being replaced in the United States by instruments such as the *dichotomous sampler*, which collects two samples, one of only fine particles and one of the larger particles smaller than 15 μm. In Europe a method known as "British Smoke" is used. This technique collects only fine particles, but it measures the collected particles by their light reflectance. The results are expressed in $\mu g/m^3$, but they have little relationship to a mass concentration in the air. A similar device is used in Canada and the United States, but its measurements are reported in "coefficient-of-haze" (COH) units. Conversion factors between British Smoke, COH units, and $\mu g/m^3$ concentrations have been derived; but they are highly dependent

Appendix A

on local conditions and are not necessarily consistent even from season to season in the same location. Large particles may be measured by collecting the particles that settle out of the air into an open-topped "dustfall jar."

Visibility measurements can be made with an integrating nepthelometer, which directly measures the amount of light scattering by the particles in the air, or with a multicontrast telephotometer, which measures the contrast of a distant object with the horizon sky. Both methods require several assumptions in order to use their measurements to calculate the meteorological range.

Health Effects. Particles larger than about 12 μm tend to be captured in the nose. Fine particles can penetrate to the deep lung, where they may remain for very long periods of time and may interfere with the foreign-matter-removal mechanisms in the lung. Because the health effects of particles are related primarily to the particles less than 12 μm, the relationship of health-effects measures to particulate-matter measurements taken by the high-volume method are tenuous. Unfortunately, the few studies that have related "respirable particles" to health effects have used neither a consistent measurement technique nor a consistent definition of what size particles are "respirable."

A comparative study of school children in two cities with low sulfur dioxide levels found a greater incidence of respiratory illness and decreased lung function in children in the more-polluted city, with an average concentration of particles of about 130 $\mu g/m^3$ (measured by the high-volume method) and a concentration of particles less than 3.5 μm of 55 $\mu g/m^3$. Most other health-effects studies have been made where both sulfur oxides and particles were present. A study of chronic-bronchitis patients in Great Britain found increased symptoms reported on days when particulate-matter concentrations were above 250 $\mu g/m^3$ (measured by the British Smoke method) and sulfur-dioxide concentrations exceeded 0.19 ppm. A long-term study of children in four English cities found that respiratory illness and impaired lung function decreased as the exposure of the children to particles and sulfur dioxide decreased. A Canadian study observed a strong relation between admissions to four hospitals for respiratory illness and a weighted sum of sulfur-dioxide and COH measurement, with no threshold of effects.

Environmental Effects. Particulate matter on the surfaces of plants may reduce their growth by interfering with photosynthesis. Damage also may result from the chemical properties of the particles. The soiling of building materials increases with higher concentrations of particulate matter; damage to the material often occurs during attempts to clean the surface. Maintenance intervals of paint were found to decrease as the concentration of particulate matter increased. Public attitudes toward air pollution have been found to be related to the amount of particulate matter that settles to the ground. One-third of the public was annoyed where settled particulate matter exceeded 10 g/m² per month.

Fine particles scatter light in the air, obscuring distant objects in a gray haze. Even when objects can still be seen, they will have lost some of their color and will appear to be less "crisp." This can occur with as little as 1.5 µg/m³ of fine particles present; at 8–12 µg/m³ of fine particles, the visibility degradation will be quite obvious.

Further Information. National Research Council (1979a).

Pesticides and Herbicides

Physical Form. Pesticides and herbicides will be observed as airborne droplets or gases.

Major Sources. As much as 35 percent of an applied insecticide or herbicide may evaporate from the ground and spread downwind. Approximately 3–5 percent of the airborne droplets will drift away from the intended application area. Four-percent drift could total 200 tons in a typical application.

Ambient Concentrations. Pesticides and herbicides are used on a massive scale in U.S. and Canadian agriculture. Problems are most likely to arise in small towns near agricultural areas or in areas where orchards or other sensitive crops are grown near application areas. In one agricultural area a monthly ambient average of 310 µg/m³ of an herbicide was measured. Analysis indicated the spray may have drifted as much as 30 km.

Health Effects. The complexity and number of pesticides and herbicides in use precludes a discussion of the health effects of each chemical or even of each class of chemicals. The most widely used are the organic phosphates, the carbamates, and the chlorinated hydrocarbons. Inhalation of organic phosphates can cause headaches, nausea, and labored breathing.

Environmental Effects. Chlorinated hydrocarbons and other pesticides have had numerous effects on the environment, substantially reducing the populations of some animals and disrupting ecosystems.

Further Information. Edwards (1973), U.S. National Institute for Occupational Safety and Health (1978).

Appendix A

Polycyclic Organic Matter

Physical Form. Polycyclic organic matter (POM) occurs as a gas, as fine particles, and condensed on the surface of other particles. The organic compounds causing the most concern are those containing two or more benzene rings in their molecular structure, including polynuclear aromatic hydrocarbons (such as benzo(a)pyrene), various arenes, the polychloro compounds, and some pesticides and herbicides.

Major Sources. The primary sources of POMs are wood- and coal-fired residential fireplaces, motorcycles and diesel motor vehicles, coal-refuse burning, coke production for the steel industry, and forest-burning activities. POM emissions can be controlled by thermal incineration or high-energy wet scrubbers.

Ambient Concentrations. The highest concentrations of POMs are in those areas where coal is still used as a household fuel and in areas near industrial sources. Measurement of the various organic pollutants generally is done with a form of chromatography followed by spectrophotometry or even mass spectrometry.

Health Effects. Benzo(a)pyrene has been identified as the principal cancer-causing agent in soot. About two dozen other polycyclic organic compounds also have been shown to cause cancer in animals. Studies of workers exposed to coke-oven emissions, which are high in benzo(a)pyrene, have clearly established a much higher risk of lung and kidney cancer and a higher incidence of other respiratory diseases among these workers. POMs also have been shown to be involved in the formation of breast cancers and to cause injury to the bone marrow, destroying the ability to form blood cells.

Further Information. National Research Council (1972).

Radioactive Matter

Physical Form. Either gases or particles may be radioactive. The radioactive isotopes of hydrogen, iodine, krypton, radium, radon, strontium, plutonium, and xenon are of most concern.

Major Sources. The principal sources of airborne radioactive matter are atmospheric nuclear testing, tailings from uranium mining, and reactor-fuel-processing facilities. Even with only a small amount of radioactive matter in coal, a coal-fired power plant can be a larger source of radioactive-matter air pollution than a nuclear power plant.

Ambient Concentrations. Many radioactive compounds are very longlived and very mobile. Therefore, once released into the atmosphere, radioactive materials spread readily throughout the environment.

Health Effects. Inhalation of radioactive gases or soluble particles can increase the body's level of radioactivity and increase the risk of leukemia and of bone and other cancers. Inhalation of certain radioactive-isotope particles has been shown to cause lung cancer and other lung injury in animals, and is presumed to have the same effect in humans.

Further Information. National Research Council (1979b).

Sulfur Oxides

Physical Form. Sulfur dioxide is a colorless, irritating gas. It readily combines with water to become sulfurous acid. Either in water or in the air, it can be oxidized easily to sulfuric acid or sulfates. Although technically incorrect, the term *sulfate* is used to include the sulfurous-acid and sulfite forms. Sulfate may occur as droplets or dry particles.

Major Sources. In Canada and the United States more than 50 percent of sulfur-dioxide emissions are from electric-power plants. The primary-metals industries account for another 20 percent of the emissions. Industrial boilers and manufacturing processes are the next-most-important sources. Exhaust gas with more than 3-percent sulfur dioxide can be controlled in a sulfuric-acid plant. Less-concentrated effluent can be controlled by a wet or dry scrubber. Several designs of wet scrubbers produce sulfur, concentrated sulfur dioxide, superphosphate fertilizers, or other useable byproducts. High-energy wet scrubbers and dry scrubbers also can be used to simultaneously control particulate-matter emissions. Residential-fuel combustion also is a significant source of sulfur dioxide. These emissions can be reduced by controlling the amount of sulfur in the coal or fuel oil. Natural sources of sulfur probably account for no more than 1–2 percent of total emissions over land.

Ambient Concentrations. Sulfur dioxide occurs at moderate concentrations in or near most urban areas. The highest concentrations occur near stationary sources. Ozone and other reactive substances in the air convert sulfur dioxide to sulfate at a rate of 1–16 percent per hour. Higher concentrations of sulfates may be observed where there is photochemical smog. Because they are so very small, sulfate particles remain suspended in the air for long periods and travel hundreds of kilometers. For this reason, sulfate concentrations tend to be constant over a large area.

Measurements of sulfur dioxide are made by continuous-gas-phase chemiluminescence or by drawing an air sample through a chemical solution (pararosaniline) and analyzing the sample by spectrophotometry. Concentrations above 200 g/m^3 measured by the pararosaniline technique will be biased low. Some summertime values measured by this method prior to 1976 also will be biased low because of adverse temperature effects on the chemicals. Sulfates are measured by collecting particles on a glass-fiber filter and analyzing by wet-chemistry spectrophotometry. Evidence exists that sulfur dioxide may convert to sulfate directly on the glass-fiber filters, overstating the amount of sulfate present in the air.

Sulfur oxides have been measured by a passive technique called *sulfation*, which measures hydrogen sulfide and sulfate particulate as well as sulfur dioxide. The sulfation "candles" or "plates" also react with other pollutants. The rates of the reactions vary with temperature and wind speed. Recommended conversion factors from sulfation to equivalent sulfur-dioxide concentrations differ by a factor of five. Even in the same location, the conversion factor can vary significantly from month to month.

Health Effects. Because of its high solubility, a substantial amount of the sulfur dioxide inhaled through the nose will be absorbed by the moisture in the nose and throat. The absorbed sulfur dioxide is rapidly taken up by the blood. The respiratory tract is somewhat protected by ammonia produced in the mouth. Brief exposures (less than one hour) at concentrations of 0.75-1 ppm of sulfur dioxide have produced changes in lung function in healthy young men and substantial changes among exercising youths with asthma.

Epidemiological studies of sulfur-dioxide and sulfate pollution are complicated by the presence of other pollutants, such as particulate matter and ozone, in the air of the study cities. Numerous analyses of short-term, high-pollution episodes or long-term mortality and morbidity rates have found a correspondence between death or illness and pollution concentrations, even after statistical corrections for temperature, weather, influenza epidemics, and so forth. Persons over 55 years of age with chronic bronchitis and emphysema are believed to be the group most vulnerable to sulfur-dioxide pollution. Some epidemiological studies have related increased sulfate levels to decreased lung function and increased illness. Four massive studies conducted by the EPA (the CHESS program) attempted to quantify the critical exposure levels for adverse health effects from sulfates. These studies have been the subject of severe criticism, although some of the results remain useful. More-recent laboratory studies have observed little, if any, adverse effects from sulfates.

Environmental Effects. Substantial yield reductions in alfalfa and other crops have been reported from short-term exposures to 0.50 ppm of sulfur dioxide and from simultaneous short-term exposures to 0.10 ppm of sulfur dioxide and 0.10 ppm of ozone. Long-term exposure of pasture grasses to 0.02 ppm of sulfur dioxide resulted in a yield loss of 24 percent. Similarly, a long-term exposure of wheat to about 0.02 ppm of sulfur dioxide caused a 15-percent yield loss. The trees most sensitive to sulfur-dioxide exposures are larch, pine, birch, and Douglas fir. Moderate injury to birch has been observed where annual average sulfur-dioxide levels were 0.03 ppm. Sulfur dioxide at annual average concentrations of less than 0.02 ppm has resulted in significant reductions in many lichen and moss populations.

Sulfur dioxide is an important source of increased acidity in rainfall, which may result in increased acidity of lakes and streams. In Scandinavia, entire biologic communities of plants, small food animals, and fish have disappeared from whole lakes and river systems. Similar effects are now being reported in eastern Canada and the northeastern United States.

Sulfur dioxide and acid rain can cause extensive damage to decorative and structural stone. Where relative humidity is moderate to high, sulfur dioxide can damage steel and other metals. Oil-based paint also is damaged by sulfur dioxide.

Further information. U.S. Environmental Protection Agency (1981).

Appendix B: Air-Pollution-Control Equipment

This appendix summarizes information on the most frequently used "tail-end" air-pollution control devices.

Particulate matter suspended in a gas exhaust stream is controlled by physically separating the particles from the gas stream. The *settling chamber* and *mechanical* (also called *cyclone*) *collector* use gravitational and inertial forces to separate the particles. The *electrostatic precipitator* uses electrical forces to separate the particles. The *fabric filter* (also called a *baghouse*) simply filters the particles out of the air. Finally, the *wet scrubber* collects the particles on water droplets, which must then be separated from the gas stream.

Gases mixed with air may be controlled by dissolving the gas into a liquid in a *wet scrubber* or *absorbtion tower*. An *adsorber* (note the slightly different spelling) removes the pollutant by selectively adsorbing it onto a special material, such as activated charcoal. Finally, a gaseous pollutant may be converted to a less harmful form by burning it in a *catalytic-oxidation* or *thermal incinerator*.

As might be expected, each control device has its advantages in specific situations, and different industries will tend to emphasize different ones. Table B-1 lists the average annual expenditures in the United States on commercial air-pollution-control devices. The data in this table represent only the purchase prices of the devices and include only the sales from a sample of manufacturers. Devices that are assembled from components or built entirely by custom contractors or the ultimate user are not included.

Mechanical Collectors

The most important consideration in removing particles from an exhaust gas is the size of the particles. The larger particles are more massive and will have much less tendency to be carried along by the gas stream. If they are emitted to the ambient air, they will quickly fall to the ground. Most such large particles need to be eliminated from an exhaust gas simply to save wear and tear on the ducting, fans, and other equipment. Often they will be removed in an enlarged duct or bin where the air is slowed down long enough that the particles can fall to the bottom and be removed.

Particles less than about 40 μm (micrometers) in diameter (a human hair is about 100 μm thick) are more difficult to remove. In the cyclone col-

Table B-1
Average Annual U.S. Purchases of Air-Pollution-Control Equipment by Industry
(millions of dollars)

Sector	Electrostatic Precipitators	Fabric Filters	Mechanical Collectors	Particulate Scrubbers	Gas Scrubbers	Adsorbers	Catalytic Oxidation	Thermal Incinerators
Cement manufacturing	9	12	4	<1	—	—	—	—
Iron and steel	20	31	3	6	1	—	—	<1
Nonferrous metals	6	14	1	4	1	—	—	<1
Grain handling and milling	—	7	<1	—	—	—	—	—
Pulp and paper	22	7	1	3	1	1	—	1
Chemicals	5	12	2	8	5	2	<1	10
Petroleum	3	1	1	4	5	—	—	3
Industrial steam boilers	5	8	1	2	9	—	—	—
Electric utilities	200	40	3	10	84	—	—	—

Source: Estimated from data in U.S. Department of Commerce, Bureau of the Census, Annual, 1975–1979, *Selected Industrial Air Pollution Control Equipment* (MA-35J); and Robert W. McIlvaine and Marilyn Ardell, *Research and Development and Cost Projections for Air Pollution Control Equipment*, EPA 600/7-78-092 (U.S. Environmental Protection Agency, 1978).

Note: In millions of 1978 U.S. dollars.

Appendix B

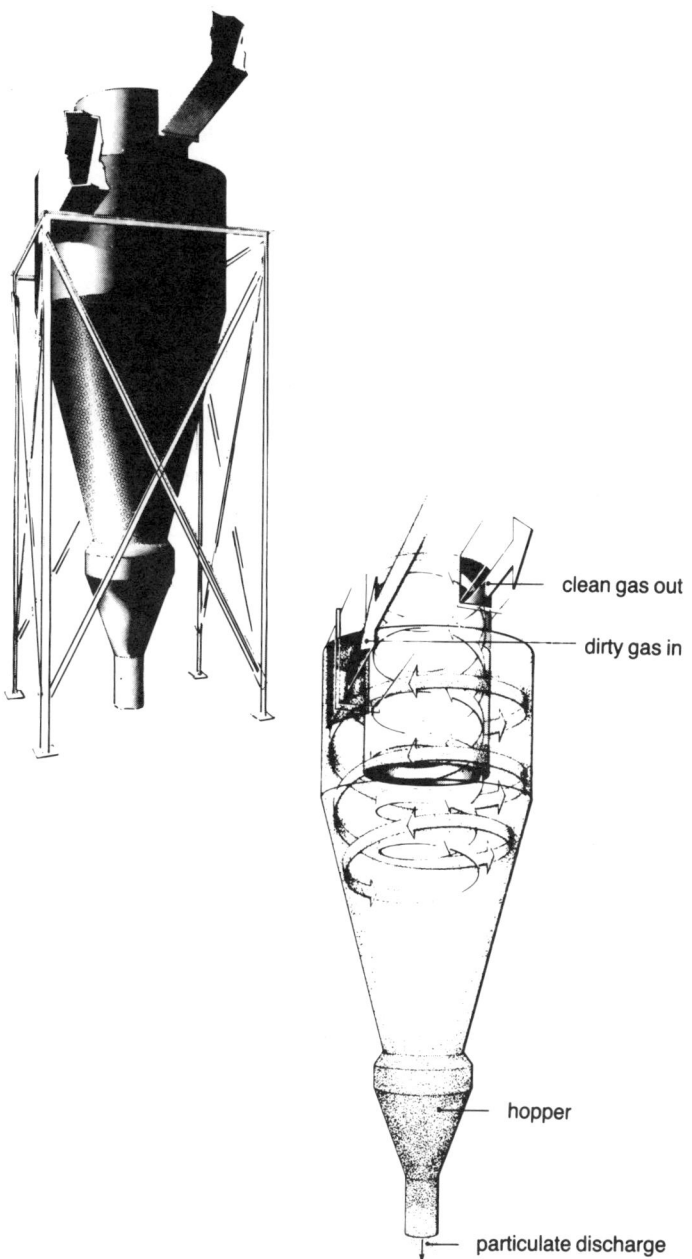

Source: American Lung Association, *Controlling Air Pollution* (New York: American Lung Association, 1974), reprinted by permission.

Figure B-1. Cyclone Mechanical Collector for Particulate Matter lector, shown in figure B-1, the inertia of the particles is used to separate them from the exhaust gas. The gas stream enters the cylindrical chamber along a side wall, which initiates a swirling motion that becomes more acute as the gas stream moves down into the conical section. Near the bottom, the gas flow reverses and climbs upward in a tight spiral flow to exit through a centrally located exhaust duct. The smaller particles are able to follow the gas molecules in the air stream, but the inertia of larger particles makes it more difficult for them to follow the turns. They move to the outer wall of the collector, where they slide down the wall and into the hopper. A well-designed single cyclone can remove most of the particles larger than about 15 μm in diameter.

Because the cyclone removes particles of different size with different efficiencies, it is necessary to know the mass of particulate matter in each size range in order to determine just how much will be removed by a specific cyclone or, conversely, to determine what size of cyclone would be needed to achieve a required overall efficiency. Several empirical relations have been developed to estimate the size-specific efficiency of a cyclone if geometric dimensions and the velocity of the gas are known. In general, the efficiency of the cyclone increases as the diameter decreases and the length increases. Up to a point the efficiency increases as the gas velocity increases, but the pressure drop increases with the square of the velocity. Because it is the pressure drop that must be overcome with fan power, greater efficiency will mean greater operating costs.

One approach to higher efficiency is to place a large number of small-diameter cyclones in a single unit. This is called a multicyclone, or multiclone. Such units can remove most of the particles larger than about 8 μm at a reasonable pressure drop. However, the increased complexity of the unit does increase the maintenance required.

Electrostatic Precipitators

Electrical forces also can be used to remove particles from the exhaust gases. In an electrostatic precipitator, shown in figure B-2, a strong electric field is established between a charged wire and a large metal plate. The air near the wire becomes ionized, and the free electric charges attach to the particles in the gas stream. With this charge of static electricity, the particle is attracted to the oppositely charged collecting plate. Once on the plate, the particles build up into a dust layer, which is periodically knocked free and drops into the hoppers.

Electrostatic precipitators can achieve total collection efficiencies of 99.9 percent and better. Particles as small as 1 μm can be collected with better than 90-percent efficiency. The efficiency is determined by the time the particles spend in the precipitator, the strength of the electric field, and the

Appendix B

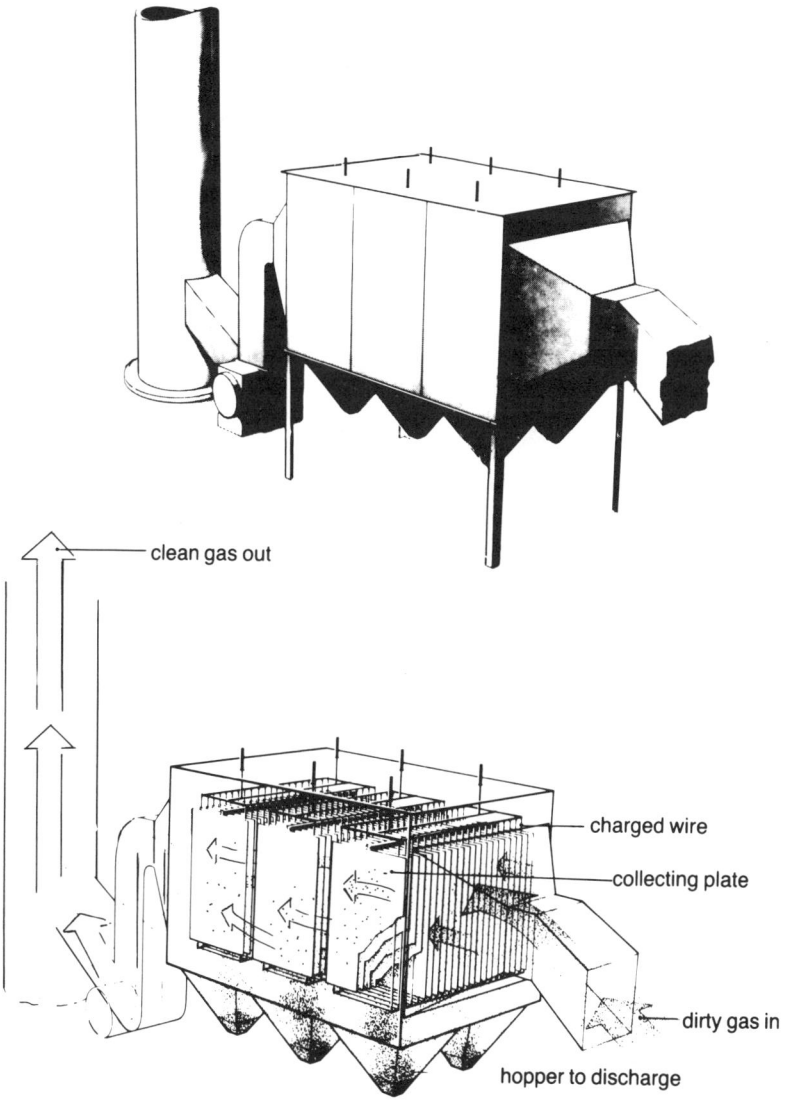

Source: American Lung Association, *Controlling Air Pollution* (New York: American Lung Association, 1974), reprinted by permission.

Figure B-2. Electrostatic Precipitator for Particulate Matter

electrical properties of the particles. In some cases the electrical properties of the particles can be modified by adding chemicals such as sulfur trioxide to the gas, or by operating the precipitator at higher temperatures.

Some electrostatic precipitators use a continually flowing, thin water film to remove the particles from the collecting plates to the hoppers. Some precipitators are designed with the collecting plate as a tube surrounding the charged wire.

Because there is little obstruction to the gas flow, electrostatic precipitators have very little pressure drop. When they are operated properly, they require very little electrical current, even though the voltages are large. Thus the operating cost of the electrostatic precipitator is the lowest of all the common tail-end devices.

Fabric Filters

The fabric filter, shown in figure B-3, collects even the very small particles with close to 100-percent efficiency. The fabric is sewn into bags (often as long, narrow cylinders), which are hung in an enclosure termed a *baghouse*. The particle-laden air enters the bottom of the baghouse. It passes up into and through the fabric bags, where the particles are removed by the fabric and the dust cake that forms on the inside of the bags. Periodically, the bags are shaken, the direction of the air flow is reversed, or a pulse of air is sent into the bag to break up the filter cake, allowing it to drop down into the hoppers.

Most shaker and reverse-air-cleaning fabric-filter systems are designed so that the cleaning operation takes place in individual sections of the baghouse in rotation. Because each section is isolated from the incoming exhaust-gas stream as it is cleaned, the baghouse design must include enough extra bags to allow one section to be out of service at any time. Most pulse-jet baghouses are designed so that the bags remain in use while they are being cleaned and extra bags are not needed.

The number of bags required is determined by the maximum acceptable pressure drop across the dust cake and the bags. In general this will be a function of the type of dust and the type of cloth. The pressure drop across the filter cake can be reduced by as much as a factor of four with some bags and dusts by putting an electric charge on the particles.

Various types of fabrics are used in fabric filters. The important considerations are cost, operating temperature, and durability, as well as filtering behavior. Cotton, nylon, and a few other fibers can be used only at temperatures below 200 °F. Fiberglass bags can be used at temperatures up to 550 °F. Cotton is quite durable, for example, whereas fiberglass must be treated gently.

The operating costs of the fabric filter are relatively high because the pressure drop across the bag and the filter cake has to be overcome with fan power. Because the repeated cleaning wears out the bags, they must be replaced every few years. Intermittent, unexpected bag failure can further increase maintenance costs.

Appendix B   221

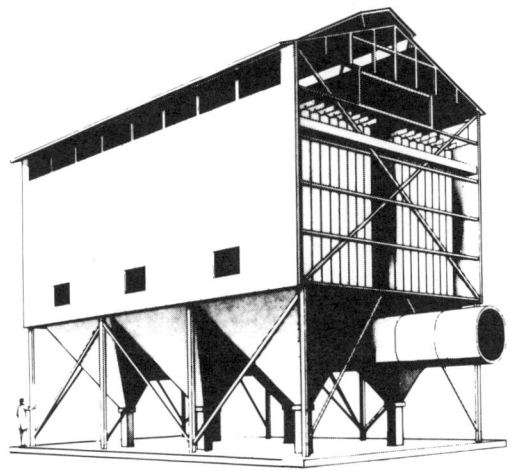

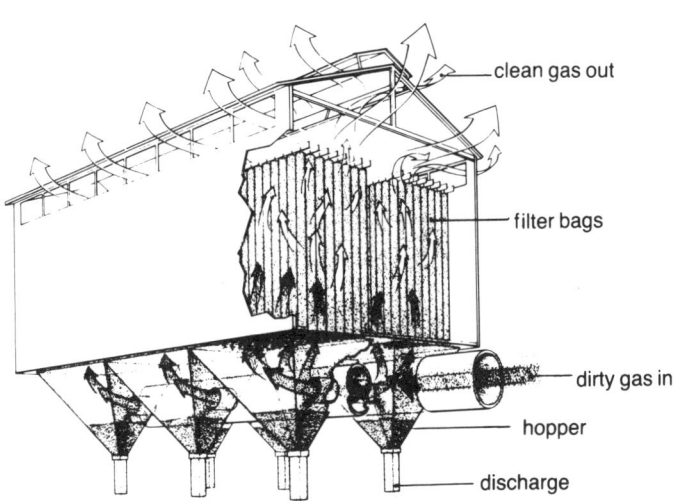

Source: American Lung Association, *Controlling Air Pollution* (New York: American Lung Association, 1974), reprinted by permission.

Figure B–3. Fabric Filters for Particulate Matter

Scrubbers

Scrubbers may be used to control either particles or gases. The simplest scrubber is a spray tower, which is commonly a vertical tower with a number of spray nozzles inside. The gas enters at the bottom of the chamber and moves countercurrent to the rain of droplets falling through the tower. Particles that run into the droplets will be removed from the air. The evaporation of the droplets will cool the air. Finally, pollutant gases in the air stream can be absorbed into the droplets, especially if the droplets are from a liquid that is chemically reactive with the pollutant (termed a *scrubbing liquor*). The design of the specific unit will determine which of these functions will be emphasized.

The *venturi scrubber*, shown in figure B-4, consists of two sections: the venturi section, which mixes the air and the liquid, and the separator or scrubber section, which removes the droplets from the gas stream. In the venturi section the gas stream is accelerated to high velocities and then mixed with the liquid at the throat, where the liquid is shattered by the gas into very small droplets, which collide with and remove the particles or absorb the pollutant gas. Several designs are used in particulate scrubbers for the separator section. A common one is similar to the cyclone described earlier.

If the scrubber is designed to remove pollutant gases, the venturi can be followed by a secondary scrubber that continues the contact between the gas and the scrubbing liquor. This may be accomplished with any of a number of designs. A frequently used design passes the gas through a bed of plastic spheres, which are kept suspended and in constant, turbulent motion by the upward flow of gases. The droplets collect on the spheres, spreading the liquor into a thin film on the surfaces, which helps speed the further removal of the pollutant gases from the air. If the venturi scrubber is designed for gas control, it generally will be preceded by a device to remove the particles. The particles that do remain will be prevented from obstructing the motion of the packed bed by its turbulance and by sprays of fresh scrubbing liquor. The particles will be drained off with the liquor. Finally, a baffle or mesh mist eliminator will remove any droplets that remain in the clean-air stream.

The cleaning efficiency of a venturi scrubber is directly related to the total pressure drop across the unit. The pressure drop is, in turn, determined by the velocity of the gas at the throat of the venturi and the flow rate of the liquid into the venturi. The optimal rate of flow for each is determined by characteristics of the particles or pollutant gas to be controlled. An efficiency of better than 99 percent is possible for particles of greater than 1 μm in diameter. Control efficiencies for pollutant gases are about 70–90 percent. The fan power required to develop the high pressures used in

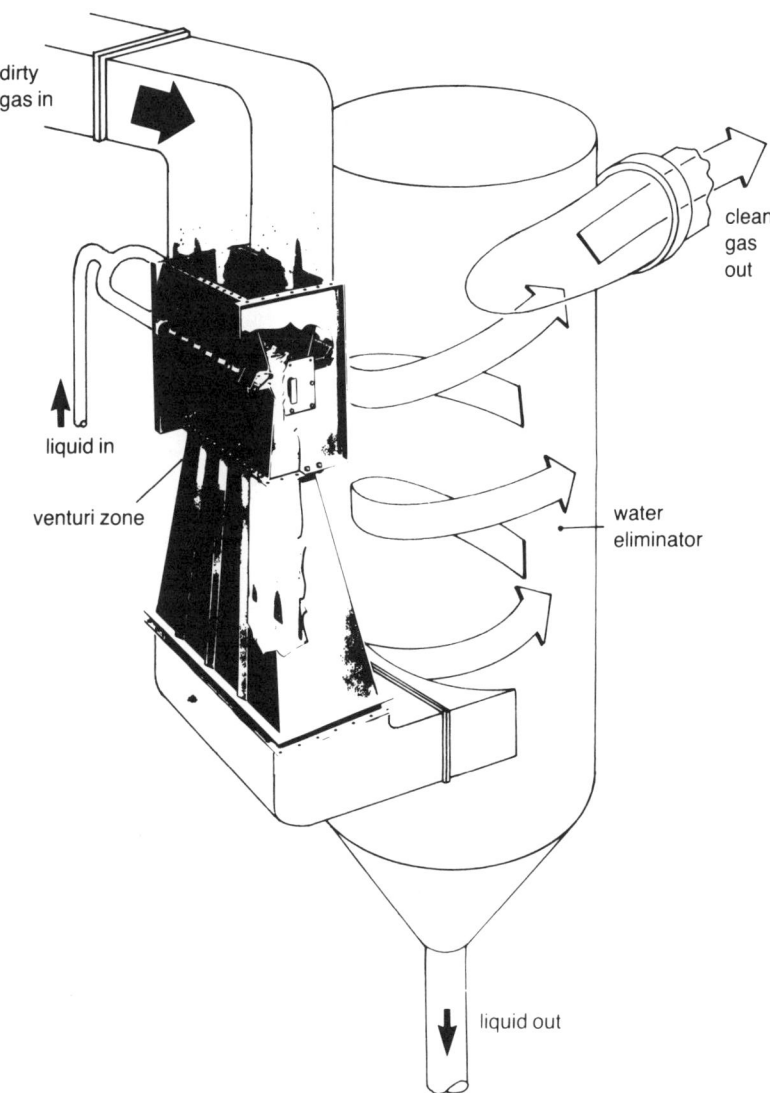

Source: Venturi section courtesy American Air Filter, reprinted by permission.
Figure B-4. Venturi Scrubber for Particulate Matter

the venturi scrubber and the tendency of the particles and/or the scrubbing liquor to deposit and foul the equipment mean high operating and maintenance costs. The collected liquid must be cleaned to avoid creating a water-pollution problem. The water added to the gas stream and the cooling of the stack gases will usually result in a white steam plume at the stack, which many members of the public may mistake for air pollution.

Tower Absorber

If there are no particles in the gas stream, an efficient device for removing pollutant gases is the tower absorber, shown in figure B-5. These devices are also called *scrubbers* by many people. The particular design shown in figure B-5 is the packed tower, which uses specially shaped fill material to provide a contact bed for the liquid and the gas. The fill material (called *packing*) is designed to create a loosely fitting maze of passages for the air stream and a maximum surface area for the thin liquid film. An alternative design is the tray or plate tower in which the liquid is held on several trays while the gas bubbles up through holes in the trays.

The liquid absorbent is introduced at the top of the tower and slowly moves downward through the packing or from one tray to the next. The gas stream enters at the bottom of the tower and moves up, with the force of the air slowing the fall of the liquid. This creates the maximum opportunity for the pollutants to transfer from the air to the liquid. The efficiency of the tower depends on the liquid flow rate, the gas velocity through the tower, and the height of the contact bed. The pressure drop will be determined by the particular type of packing used or the depth of the liquid in the trays. As might be expected, designs that are more efficient in removing the pollutants also will have a higher pressure drop.

Adsorber

In an adsorption unit, the pollutant gas is collected on the surfaces of a special, porous solid material (for example, activated charcoal). The solid material and the collected pollutant can be removed from the adsorber and discarded, or the solid adsorbent may be renewed. In most instances the renewal process can be carried out in such a way that a concentrated stream of the pollutant gas is produced, making it useful as a byproduct. The physical design of the adsorption unit is quite similar to that of the absorption tower, but without any liquid. It often will be designed for horizontal installation.

Incinerator

Gaseous pollutants also may be controlled by burning in an incinerator, which in many instances will leave little more than carbon dioxide and water

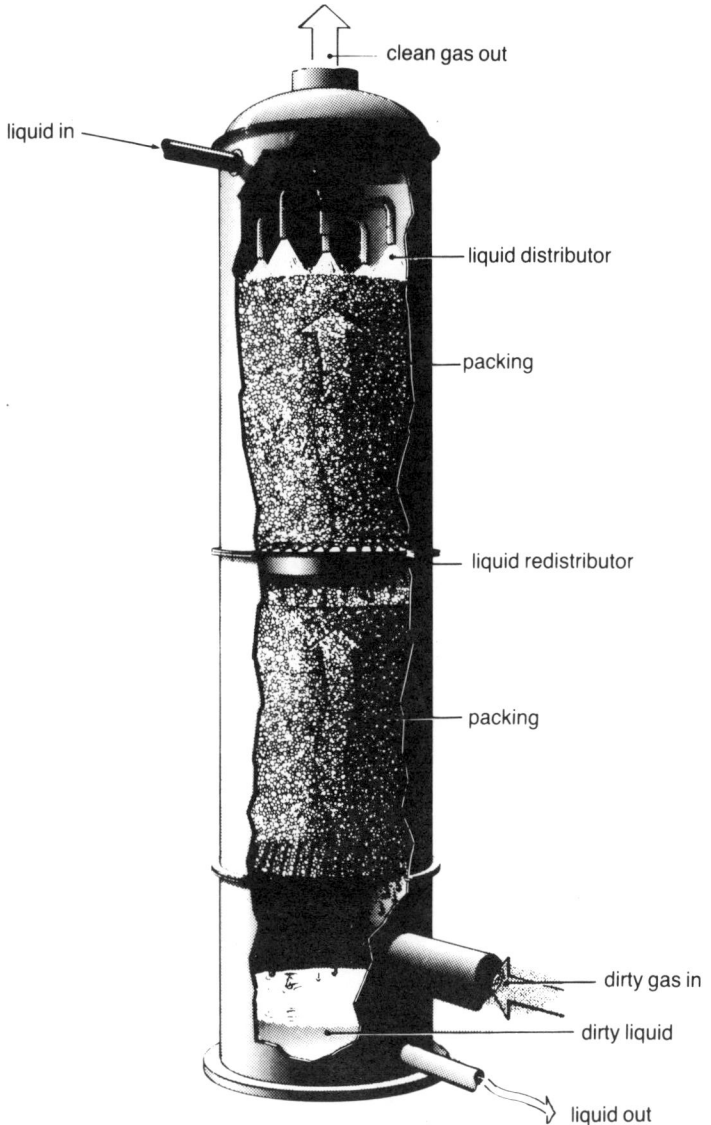

Source: American Lung Association, *Controlling Air Pollution* (New York: American Lung Association, 1974), reprinted by permission.

Figure B–5. Packed-Tower Absorber for Gases

in the air stream. In a thermal incinerator, the polluted gas stream is supplied directly to an open flame. With a sufficiently high temperature and a long-enough residence time in the flame, the pollutant will be completely oxidized. Certain catalytic materials, such as platinum, can ignite and oxidize the pollutant gases without an open flame or additional fuel. The catalyst is not used up in the reaction, although it can be fouled by particles in the gas stream. The burning of a pollutant gas in either a catalytic-oxidation or thermal incinerator will produce heat that can be used elsewhere in the process. Installing a heat exchanger will make the operation of an incinerator much more economically efficient.

Support Equipment

In addition to the pollution-control device, an air-pollution-control system may include:

Pollutant-Capture Equipment. For many processes, the exhaust gases can be vented directly to a duct that will lead to the control equipment. For others, it is necessary to capture the pollutant from a source that is open to the atmosphere. This is normally accomplished with an exhaust hood, which also will draw in air and lower the concentration and temperature of the process gases.

Gas Preconditioning. Large particles are removed in settling chambers and cyclones to reduce wear on ducts, fans, and other equipment. Cooling the gases in a heat exchanger or spray tower reduces the volume and the temperature of the gases to be handled. Insulation may be necessary to ensure that the gases do not get too cool and lose buoyancy, which would mean that they would have to be reheated at the stack.

Fans. Fans are needed to move the gas along and to overcome the resistance of the ducting and the pressure drop across the control equipment.

Auxiliary Equipment and Ducting. If the control program requires materials to be added to the exhaust gases, such as a scrubbing liquor, these must be mixed, stored, conveyed, pumped, and so forth. In very hot environments, as in the metals industries, hoods and ducts may have to be water cooled. If corrosive gases are handled, special metals may be required. If water is used as a coolant, a cooling tower may be necessary.

Waste Removal and Treatment Equipment. Collected particulate matter must be removed from the equipment and transported to a disposal site or prepared for sale. Liquids and slurries must be cleaned before disposal.

Stack. A new stack or reconstruction of the existing stack may be necessary. Emission-monitoring equipment may be required at the stack.

Appendix B

McGlammery et al. (1975) provide an example of the number of different items that could be included in a pollution-control program, although the example of the venturi scrubber that they describe is probably an extreme case.

Further description of air-pollution-control equipment is available from a number of sources. The *Air Pollution Engineering Manual* (Danielson 1973) provides detailed descriptions and calculation procedures for auxiliary equipment such as exhaust hoods and fans, as well as for control devices. A number of textbooks and comprehensive handbooks on air pollution also include articles on the various kinds of control devices (for example, Stern 1977b).

Appendix C: Order-of-Magnitude Cost Functions

The Cost of Clean Air and the Strategic Environmental Assessment System (SEAS) cost functions are presented in this appendix. Both the capital and operating cost functions are of the form

$$\text{Control Cost} = \alpha(\text{Size Variable})^\beta \qquad (C.1)$$

where the coefficients α and β are listed in table C-1 for various types of processes and control techniques. These equations represent the total cost of installing and operating the control equipment. The annual operating expenses do not include capital charges, depreciation, or similar factors. In most cases, the size variable is the quantity of product manufactured, although for surface coatings it is the quantity of coating compound consumed.

The industrial sectors are divided by specific types of plant, pollution source, production process, or control technique. Where more than one pollutant is listed or more than one pollution source is located in the same plant, it is necessary to add them together to obtain the total cost. If more than one production process or control technique is listed, the correct one must be matched to the plant being evaluated.

The costs estimated by these cost functions do not include expenditures that might be termed "good engineering practice," that is, controls installed for recovery of raw materials, protection of process equipment, or avoidance of a public nuisance. The projected level of control is either a typical requirement of present U.S. state regulations or the requirement of a U.S. New Source Performance Standard (NSPS), a national emission regulation applied only to new or substantially modified facilities. The NSPS requirement often will be more stringent than the state regulations, but since it is applied to new plants it may be less expensive to implement.

Both the Cost of Clean Air and SEAS projects are ongoing and are continually updating their cost functions. The functions from the Cost of Clean Air are more recent than the SEAS functions for most sectors and are the primary ones reported here. Additional details on each of the listed sectors are provided in the Cost of Clean Air report (U.S. Environmental Protection Agency 1979c).

Table C-1
Cost of Clean Air and SEAS Cost Functions

Process/Control	Size Variable	Pollutant	Type	Capital Cost α	Capital Cost β	Operating Cost α	Operating Cost β
Energy industries							
Commercial coal boilers							
Pulv.-fuel switch	MBTU/hr	SO_x/PM		2.16E3	0.77	7.17	1.00
Stoker-fuel switch	MBTU/hr	SO_x/PM		5.31E3	0.65	5.13	1.00
Industrial coal boilers							
Scrubber	MBTU/hr	SO_x		7.56E4	0.60	9.54	0.60
Pulv.-ESP	MBTU/hr	PM		4.39E4	1.00	26.6	1.00
Stoker-ESP	MBTU/hr	PM		3.55E4	1.00	21.5	1.00
Natural-gas processing							
Claus plant	T(S)/day	SO_x		1.24E5	0.65	−1.05E4[d]	1.05E6[d]
Claus tailgas	T(S)/day	SO_x	NS[a]	9.59E4	0.65	2.00E4	0.53
			[b]	9.59E4	0.65	1.14E4	0.68
Petroleum refining							
Crude storage	Mbbl	HC	[c]	2.45E3	0.57	−3.91E2	0.57
			NS [c]	3.05E3	0.57	−3.77E2	0.57
Gas storage	Mbbl	HC	[c]	1.01E3	0.57	−3.86E2	0.57
			NS [c]	1.26E3	0.57	−3.86E2	0.57
Jet-fuel storage	Mbbl	HC	[c]	2.54E3	0.57	−2.15E2	0.57
			NS [c]	3.12E3	0.57	−2.06E2	0.57
Catalytic cracker	kbbl/day	PM	[c]	1.68E4	0.42	6.07E2	0.52
			NS [c]	1.40E4	0.42	6.07E2	0.52
	kbbl/day	CO	[c]	6.92E5	0.34	−6.17E4	0.34
			NS [c]	5.77E5	0.34	0.69	1.20
Fuel-gas desulf.	T(S)/yr	SO_x	[c]	3.37E5	0.61	4.54E4	0.72
			NS [c]	4.28E5	0.51	4.58E4	0.73

Appendix C

Category	Process	Units						
Coal cleaning		kT/yr	PM		2.12E3	0.87	2.09E2	1.05
Chemicals industries								
Phtalic anhydride	Incinerator	Mlb/yr	HC/CO		5.30E4	0.60	1.07E4	0.99
Formaldehyde	Silver catalyst	Mlb/yr	HC/CO		9.67E3	0.49	4.08E2	0.90
	Metanol oxidation	Mlb/yr	HC/CO		1.24E4	0.45	5.72E2	0.87
Acrylonitrile	Incinerator	Mlb/yr	HC/CO		2.50E4	0.60	3.32E3	1.00
Ethylene oxide-O_2	Incinerator	Mlb/yr	HC		5.52E4	0.60	2.94E2	0.98
Ethylene oxide-air	Incinerator	Mlb/yr	HC		2.06E4	0.60	2.54E3	0.98
Vinyl chloride	Balanced-incinerator	Mlb/yr	VC	NH	1.02E6	0.00	1.78E5	0.23
	EDC-incinerator	Mlb/yr	VC	NH	1.06E4	0.60	1.06E5	0.20
	VCM-incinerator	Mlb/yr	VC	NH	3.49E4	0.60	2.10E5	0.20
	PVC-incinerator	Mlb/yr	VC	NH	4.41E5	0.51	6.29E5	0.09
Nitric acid	Cat. reduction	T/day	NO_x		1.81E4	0.60	1.36E3	0.60
				NS	1.15E4	0.60	8.66E2	0.60
Sulfuric acid	Sulfur burning	T/day	SO_x		5.95E4	0.60	3.25E3	0.71
				NS	1.76E4	0.59	6.84E2	0.84
	Wet gas	T/day	SO_x		1.13E4	0.61	7.91E3	0.64
				NS	3.80E4	0.58	1.60E3	0.74
Phosphate fertilizer	Granulated	kT/yr	PM		1.28E4	0.39	4.05E2	0.69
	Diammonium	kT/yr	PM		1.78E4	0.67	3.21E3	0.74
	Normal super	kT/yr	PM		3.32E3	0.78	5.34E2	0.87
	Triple super	kT/yr	PM		3.40E4	0.69	6.44E3	0.79
Nonfertilizer phosphorous	Phosphorous metal	kT/yr	F		1.48E4	0.72	5.46E2	0.68

Table C-1 continued.

	Process/Control	Size Variable	Pollutant	Type	Capital Cost		Operating Cost	
					α	β	α	β
	DF phosphates	kT/yr	F		1.01E4	0.65	6.30E2	0.67
	Dicalcium phos.	kT/yr	F		2.22E4	0.65	1.51E3	0.63
Chlor-alkali	Mercury cell	Mg/day	Hg	NH	2.83E4	0.60	2.66E3	0.80
Metals industries								
Sintering steel	Fabric filter	Mg/yr	PM	c	1.40E3	0.60	3.53	0.80
	Scrubber	Mg/yr	PM	NS c	1.37E3	0.60	3.53	0.80
		Mg/yr	PM	c	9.38E2	0.60	10.6	0.80
	ESP	Mg/yr	PM	NS c	8.95E2	0.60	10.6	0.80
		Mg/yr	PM	c	2.20E2	0.60	3.18	0.80
		Mg/yr	PM	NS c	2.90E2	0.60	3.18	0.80
Coke ovens	Gas desulf.	T/yr	SO_x	c	77.2	0.78	58.0	0.65
	Charging car	T/yr	PM	c	71.6	0.81	0.69	0.81
	Collection	T/yr	PM	c	17.0	0.91	0.52	0.94
Iron and steel	Open hearth	kT/yr	PM	c	4.53E6	0.79	2.75E3	0.65
	BOF–fabric	kT/yr	PM	c	3.62E6	0.73	8.17E2	0.90
	BOF–scrubber	kT/yr	PM	c	1.14E4	0.71	2.93E2	1.06
				NS c	1.61E4	0.71	3.38E2	1.06
	Electric arc	kT/yr	PM	c	51.7	1.83	3.78E3	0.59
				NS c	44.1	1.83	3.78E3	0.59
Iron foundries	Scrubber	Mg/hr	PM		1.01E5	0.36	4.88E3	0.78
				NS	1.09E5	0.40	7.23E3	0.85

Appendix C

Steel foundries		Fabric filter	Mg/hr	PM		3.03E4	0.71	1.07E4	0.47
		Fabric filter	Gg/yr	PM		1.64E4	0.33	6.57E3	0.34
					NS	7.75E5	0.20	3.33E4	0.21
Ferroalloy		Fabric filter	kW	PM	c	56.8	1.01	1.43E4	0.17
		ESP	kW	PM	c	7.56E3	0.60	62.9	0.80
Primary aluminum		Prebake	kT/yr	F	NS	7.23E4	1.00	1.67E4	1.00
		Vertical stud	kT/yr	F	NS	9.53E4	1.00	3.08E4	1.00
		Horizontal stud	kT/yr	F	NS	7.56E4	1.00	2.59E4	1.00
Primary copper		With roaster	T/yr	SO_x	c	3.38E3	0.74	3.19E2	0.76
		Without roaster	T/yr	SO_x	c	3.06E2	0.89	3.71E5	0.19
Primary lead		Acid plant	T/yr	SO_x		2.55E3	0.68	89.0	0.76
Primary zinc		Acid plant	kT/yr	SO_x		8.68E5	0.68	5.67E5	0.76
Secondary aluminum		Scrubber	T/yr	PM		3.79E3	0.49	3.97E2	0.60
Secondary brass		Fabric filter	T/yr	PM		2.27E2	0.60	1.94	0.80
Secondary lead		Fabric filter	T/day	PM		8.84E3	0.82	1.49E3	0.69
Secondary zinc		Fabric filter	T/yr	PM		1.00E3	0.63	2.35E2	0.66
Quarrying and construction industries									
Cement		Wet proc.–ESP	Gg/yr	PM	NS	4.67E4	0.67	5.14E2	1.00
		Wet proc.–fabric	Gg/yr	PM	NS	3.00E3	0.91	2.27E2	1.00
		Dry Proc.–ESP	Gg/yr	PM	NS	1.16E3	0.91	8.75E2	1.00
		Dry proc.–fabric	Gg/yr	PM	NS	9.01E3	0.91	6.82E2	1.00

Table C-1 continued.

	Process/Control	Size Variable	Pollutant	Type	Capital Cost α	Capital Cost β	Operating Cost α	Operating Cost β
Clay brick and pipe	Wet scrubber	T/day	SO_x		3.60E4	0.29	5.60E2	0.70
	Incinerator	T/day	HC		2.68E4	0.48	1.34E3	0.88
Lime	Scrubber	kT/day	PM		9.22E3	0.80	6.73E3	0.24
				NS	6.14E3	0.80	6.73E3	0.24
	ESP	kT/day	PM		3.98E4	0.70	4.86E3	0.20
				NS	2.65E4	0.70	4.86E3	0.20
	Fabric filter	kT/day	PM		2.38E4	0.90	2.51E3	0.72
				NS	9.97E3	0.90	2.51E3	0.72
Asphalt	Fabric filter	Mg/hr	PM		7.00E3	0.47	2.49E2	0.87
				NS	1.14E4	0.48	5.93E2	0.66
	Scrubber	Mg/hr	PM	NS	2.21E3	0.75	3.66E2	0.86
Asbestos	Textile	Mg/yr	PM	NH^c	1.79E3	0.60	3.56E2	0.60
	Felt-misc.	Mg/yr	PM	NH^c	2.76E2	0.60	55.2	0.60
	Construction	Mg/yr	PM	NH^c	3.36E2	0.60	79.1	0.60
Other manufacturing and services								
Paint manufacture	Fabric filter	M Gal/yr	PM		9.21E3	0.60	3.81E2	0.60
	Incinerator	M Gal/yr	HC		4.88E4	0.39	7.01E3	0.91
Dry cleaning	Petrol. solvent	T/yr	HC	c	2.02E2	1.00	−19.5	1.00
	Syn. solvent	T/yr	HC	c	34.7	1.00	−15.0	1.00
	Solvent switch	T/yr	HC	c	2.02E2	1.00	−15.0	1.00

Appendix C

Printing	Incinerator	T/yr	HC	c	1.96E3	0.58	28.7	0.90
Surface coatings:									
Automobile finishing	Prime-electrodep.	Plant	HC		9.18E6	—	-1.38E5	—
	Topcoat–low solv.	Plant	HC		1.15E6	—	1.38E5	—
	Topcoat–waterbase	Plant	HC		2.30E7	—	2.87E6	—
Furniture finishing	Carbon adsorb.	scfm	HC		5.65	1.07	2.31	0.99
Coil coating	Incinerator	Gal/yr	HC		8.75	0.80	1.56	0.80
				NS	5.00	0.80	0.89	0.80
Home appliances	Prime-electrodep.	Gal/yr	HC		5.41E5	0.00	-2.84	1.00
	Prime-waterbase	Gal/yr	HC		4.33E4	0.00	0.90	1.00
	Topcoat–low solv.	Gal/yr	HC		4.33E4	0.00	-0.64	1.00
Forest and agricultural products									
Kraft pulp	Recovery furnace	D Gg/yr	SO_x	c	5.64E3	0.63	1.16E3	0.66
		D Gg/yr	PM	c	1.86E4	0.79	2.00E3	0.83
	Lime kiln	D Gg/yr	SO_x	c	5.64E3	0.63	1.16E3	0.66
		D Gg/yr	PM	c	1.17E4	0.52	2.35E3	0.52
	Boiler	D Gg/yr	SO_x	c	2.00E5	0.70	2.04E4	0.90
		D Gg/yr	PM	c	3.06E4	0.60	9.82E2	0.90
NSSC paper	Recovery furnace	D Gg/yr	PM	c	9.19E4	0.48	1.32E4	0.53
				NS c	8.29E4	0.48	9.75E3	0.53
Grain elevators	Fabric filter	Mbsl/yr	PM		1.85E5	0.42	5.81E3	0.60
				NS	7.70E5	0.44	4.38E4	0.63
Feed mills	Fabric filter	Mg/yr	PM		4.7 E3	0.40	63.8	0.54
				NS	4.10E3	0.40	1.46E4	0.01

Table C-1 continued.

Process/Control	Size Variable	Pollutant	Type	Capital Cost		Operating Cost	
				α	β	α	β
Solid-waste disposal							
Open burning	T/day	PM		8.53E3	0.79	8.86E2	1.05
Municipal incinerator	T/day	PM		1.13E4	0.76	1.88E3	0.79
Sewage-sludge incin.	T/day	PM		5.34E4	0.76	4.25E3	0.62
Indus./comm. incin.	T/yr	PM		8.05E2	0.60	26.7	1.00

Source: M. Cohen, U.S. Environmental Protection Agency, personal communication, 1980.

Notes: Total costs in 1978 U.S. dollars. Numbers written as $1.23E6$ mean 1.23×10^6. Abbreviations for units are bbl: barrel; bsl: bushel; D: dry; g: gram; gal: gallon; lb: pound; T: ton (2,000 lbs); k: 10^3; M: 10^6; G: 10^9.

SO_x Sulfur oxides.   VC   Vinyl chloride.   NS   U.S. New Source Performance
PM   Particulate matter.   Hg   Mercury.   Standard.
CO   Carbon monoxide   HC   Hydrocarbons.   NH   U.S. National Emission Standard
NO_x Nitrogen oxides.   F   Fluoride.   for Hazardous Pollutants.

[a] > 50 T/day.
[b] < 50 T/day.
[c] SEAS cost function, otherwise from Cost of Clean Air.
[d] Coefficients are for an equation of the form

Control Cost = $\alpha + \beta$(size variable)

References

Note that the following abbreviations are used:

- AER   *American Economic Review.*
- J.   *Journal (of).*
- JEEM   *Journal of Environmental Economics and Management.*
- JAPCA   *Journal of the Air Pollution Control Association.*
- JPE   *Journal of Political Economy.*
- QJE   *Quarterly Journal of Economics.*
- U.S. EPA   U.S. Environmental Protection Agency.

Acton, Jan Paul. 1973. *Evaluating Public Programs to Save Lives: The Case of Heart Attacks.* Santa Monica, Calif.: Rand Corp.(R-950-RC).

Adams, Richard M; Thanavibulchai, Narongsakdi; and Crocker, Thomas D. 1979. *A Preliminary Assessment of Air Pollution Damages for Selected Crops Within Southern California.* Methods Development for Assessing Air Pollution Control Benefits, vol. 3. U.S. EPA (EPA 600/5-79-001c).

Allais, Maurice. 1953. Le comportement de l'homme rationnel devant le risque, critique des postulats et axiomes de l'école Américaine. *Econometrica* 21:503-546.

American Lung Association. 1974. *Controlling Air Pollution.* New York: American Lung Assn. Illustrations by David Black.

Anderson, David. 1973. *Emission Factors for Trace Substances.* U.S. EPA (EPA 450/2-73-001).

Arnold, Thomas H., and Chilton, Cecil H. 1963. New index shows plant cost trends. *Chemical Engineering* 70(4):143-152.

Arrow, Kenneth J. 1950. A difficulty in the concept of social welfare. *JPE* 58:328-346.

———. 1966. Discounting and public investment criteria. In *Water Research,* ed. Allen V. Kneese and Stephen C. Smith, pp. 13-32. Baltimore: Johns Hopkins Press.

Arrow, Kenneth J., and Lind, Robert C. 1970. Uncertainty and the evaluation of public investment. *AER* 60:364-378.

Azzi, Corry F. and Cox, James C. 1973. Equity and efficiency in evaluation of public programs. *QJE* 87:495–502.

Babcock, Lyndon R., and Nagda, Niren L. 1976. *Popex: Ranking Air Pollution Sources by Population Exposure.* U.S. EPA (EPA 600/2-76-063).

Bailey, Martin J. 1979. The possibility of rational social choice in an economy. *JPE* 87:37–56.

———. 1980. *Reducing Risks to Life.* Washington, D.C.: American Enterprise Inst. for Public Policy Research.

Baram, Michael S. 1980. Cost-benefit analysis: An inadequate basis for health, safety, and environmental regulatory decision-making. *Ecology Law Quarterly* 8:473–531.

Bator, Francis M. 1957. The simple analytics of welfare maximization. *AER* 47:22–59.

Baumol, William J. 1977. *Economic Theory and Operations Analysis,* 4th ed. Englewood Cliffs, N.J.: Prentice-Hall.

Baumol, William J., and Oates, Wallace E. 1975. *The Theory of Environmental Policy.* Englewood Cliffs, N.J.: Prentice-Hall.

Beloin, Norman J. 1973. Fading of dyed fabrics exposed to air pollutants. *Textile Chemist and Colorist* 5:128–133.

Beloin, Norman J., and Haynie, Fred H. 1975. Soiling of building materials. *JAPCA* 25:399–403.

Benarie, M. 1976. Empirical dosage–distance relationships around a point source at ground level. *Atmospheric Environment* 10:163–166.

Benarie, M.; Badellon, D.; Menard, T.; and Nonat, A. 1974. The use of multiple regression equations for the assessment and short-term forecasting of urban area background pollution. In *Statistical and Mathematical Aspects of Pollution Problems,* ed. John W. Pratt, pp. 249–273. New York: Marcel Dekker.

Benedict, Harris M., and Jaksch, John A. 1979. Economic assessment of damage to vegetation. In *Methodology for the Assessment of Air Pollution Effects on Vegetation,* ed. W.W. Heck, S.V. Krupa, and S.N. Linzon, chap. 15. Pittsburgh: Air Pollution Control Assn.

Benedict, Harris M.; Miller, Clarence J.; and Smith, Jean S. 1973 *Assessment of Economic Impact of Air Pollutants on Vegetation in the United States: 1969 and 1971.* Menlo Park, Calif.: Stanford Research Inst. (EPA 650/5-78-002).

Bergson, Abram. 1938. A reformulation of certain aspects of welfare economics. *QJE* 52:310–334.

Blecker, Herbert G., and Cadman, Theodore W. 1973. *Capital and Operating Costs of Pollution Control Equipment Modules,* vol. 1. U.S. EPA (EPA RS-73-023a).

Blecker, Herbert G., and Nichols, Thomas M. 1973. *Capital and Operating*

Costs of Pollution Control Equipment Modules, vol. 2. U.S. EPA (EPA RS-73-023b).

Bloom, S.G., Rosenberg, H.S., Hissong, D.W., and Oxley, J.M. 1978. *Analysis of Variations in Costs of FGD Systems.* Palo Alto, Calif.: Electric Power Research Inst. (EPRI FP-909).

Boadway, Robin W. 1976. Integrating equity and efficiency in applied welfare economics. *QJE:* 90:541-556.

———. 1978. Public investment decision rules in a neo-classical growing economy. *International Economic Review* 19:265-287.

Booz, Allen, and Hamilton, Inc. 1970. *Study to Determine Residential Soiling Costs of Particulate Air Pollution.* Washington, D.C.: Booz, Allen, and Hamilton (PB 205 807).

Bradford, David F. 1975. Constraints on government investment opportunities and the choice of discount rate. *AER* 65:887-899.

Brookshire, David S.; d'Arge, Ralph C.; Schulze, William D.; and Thayer, Mark A. 1979. *Experiments in Valuing Non-Market Goods: A Case Study of Alternative Benefit Measures of Air Pollution Control in the South Coast Air Basin of Southern California.* Methods Development for Assessing Tradeoffs in Environmental Management, vol. 2. U.S. EPA (EPA 850/6-79-001b).

Brown, Charles. 1980. Equalizing differences in the labor market, *QJE* 94:113-134.

Browning, Edgar K. 1976. The marginal cost of public funds. *JPE* 84:283-298.

Brysson, Ralph J.; Trask, Brenda J.; Upham, James B.; and Borras, Samuel G. 1967. The effects of air pollution on exposed cotton fabrics. *JAPCA* 17:294-298.

Burchard, John K.; Rochelle, Gary T.; Schofield, William R.; and Smith, John O. 1972. Some general economic considerations of flue gas scrubbing for utilities. Paper presented to the Electrical World Conference on Sulfur in Utility Fuels (APTIC 55207).

C-E Air Preheater. 1976. *Report of Fuel Requirements, Capital Cost and Operating Expense for Catalytic Afterburners.* U.S. EPA (EPA 450/3-76-031).

Campbell, G.G.; Schurr, G.G.; Slawikowski, D.E.; and Spence, J.W. 1974. Assessing air pollution damage to coatings. *J. Paint Technology* 46(593):55-71.

Carlson, John A. 1977. Short-term interest rates as predictors of inflation: Comment. *AER* 67:470-475.

Caspari, Conrad; MacLaren, Donald; and Hobhouse, Georgina. 1980. *Supply and Demand Elasticities for Farm Products in the Member Countries of the European Community.* U.S. Department of Agriculture (PB 80-204316).

Chapman, Richard A.; Clements, Donald P.; Sparks, Leslie E.; and Abbott, James H. 1980. Cost and performance of particulate control devices for low-sulfur western coals. In *Second Symposium on the Transfer and Utilization of Particulate Control Technology,* vol. 1, pp. 1-14. U.S. EPA (EPA 600/9-80-039a).

Chipman, John S., and Moore, James C. 1978. The new welfare economics 1939-1974. *International Economic Review* 19:547-584.

Coase, Ronald. 1960. The problem of social cost. *J. Law and Economics* 3:1-44.

Cohen, Alan S. 1977. *An Economic Evaluation of Proposed Amendments to the Illinois Sulfur Dioxide Regulations R74-2, R75-5 and R76-9.* Chicago: Illinois Inst. for Environmental Quality (PB 282 390).

Cohen, Alan S.; Fishelson, Gideon; and Gardner, John L. 1974. *Residential Fuel Policy and the Environment.* Cambridge, Mass.: Ballinger.

Cooper, Barbara S., and Rice, Dorothy P. 1976. The economic cost of illness revisited. *Social Security Bulletin* 39(February):21-36.

Cran, John. 1976. Cost indices. *Engineering and Process Economics* 1:13-23.

Crittenden, P.D. and Read, D.J. 1978. The effects of air pollution on plant growth with special reference to sulfur dioxide. II. Growth studies with *Lolium Perenne* L. *New Phytologist* 80:45-62.

Crocker, Thomas D.; Schulze, William; Ben-David, Shaul; and Kneese, Allen V. 1979. *Experiments in the Economics of Air Pollution Epidemiology.* Methods Development for Assessing Air Pollution Control Benefits, vol. 1. U.S. EPA (EPA 600/5-79-001a).

Culyer, A.J. 1977. The quality of life and the limits of cost-benefit analysis. In *Public Economics and the Quality of Life,* ed. Lowden Wingo and Alan Evans, pp. 141-153. Baltimore: Johns Hopkins Univ. Press.

Danielson, John A., ed. 1973. *Air Pollution Engineering Manual,* 2d ed. U.S. EPA (AP-40).

Davis, C.R. 1972. Sulfur dioxide fumigation of soybeans: Effect on yield. *JAPCA* 22:778-780.

Davis, Donald D., and Wilhour, Raymond G. 1976. *Susceptibility of Woody Plants to Sulfur Dioxide and Photochemical Oxidants.* U.S. EPA (EPA 600/3-76-102).

DeMandel, R.E.; Robinson, Lewis H.; Fong, James S.L.; and Wada, Ronald Y. 1979. Comparisons of EPA rollback, empirical/kinetic, and physiochemical oxidant prediction relationships in the San Francisco Bay area. *JAPCA* 29:352-358.

Devitt, Timothy W., Spaite, P.; and Gibbs, L. 1979. *Population and Characteristics of Industrial/Commercial Boilers in the U.S.* U.S. EPA (EPA 600/7-79-178a).

Devitt, Timothy W.; Yerino, Lario V.; Ponder, Thomas C. Jr.; and Chatlynne, C.J. 1976. Estimating costs of flue gas desulfurization systems for utility boilers. *JAPCA* 26:204-209.

Dewees, Donald. 1974. Cost of alternative automobile emission standards. In *Costs and Benefits of Automobile Emission Control,* pp. 25-128. Air Quality and Automobile Emission Control, vol. 4. U.S. Senate, Committee on Public Works (Serial 93-24).

Dorfman, Nancy S., and Snow, Arthur. 1975. Who will pay for pollution control?—The distribution by income of the burden of the national environmental protection program, 1972-1980. *National Tax J.* 28:101-115.

Dravnieks, Andrew, and O'Neill, Hugh J. 1979. Annoyance potentials of air pollution odors. *American Industrial Hygiene Assn. J.* 40:85-95.

Edmiston, Norman G., and Bunyard, Francis L. 1970. A systematic procedure for determining the cost of controlling particulate emissions from industrial sources. *JAPCA* 20:446-452.

Edwards, Clive Arthur, ed. 1973. *Environmental Pollution by Pesticides.* London: Plenum Press.

El-Sawy, A.; Leigh, J.G.; and Trehan, R.K. 1979. A quantitative comparison of energy costing methods. *Energy Systems and Policy* 3:213-226.

Ensor, David S., and Pilat, Michael J. 1971. Calculation of smoke plume opacity from particulate air pollutant properties. *JAPCA* 21:496-501.

Ensor, David S.; Sparks, Leslie E.; and Pilat, Michael J. 1973. Light transmittance across smoke plumes downwind from point sources of aerosol emissions. *Atmospheric Environment* 7:1267-1277.

Fabrik, A.; Skarlew, R.; and Wilson, J. 1977. *Point Source Modeling.* Westlake Village, Calif.: Form and Substance.

Fama, Eugene F. 1976. Inflation uncertainty and expected returns on treasury bills. *JPE* 84:427-448.

Farber, Paul S. 1977. Capital and operating costs of particulate control equipment for coal-fired power plants. Paper presented to the Fifth National Conference on Energy and the Environment (CONF 7710101).

Farber, Paul S., and Livengood, C.D. 1979. Energy and economic impacts of pollution control equipment for coal-fired power plants: An assessment model. Paper presented to the Annual Meeting of the Air Pollution Control Assn. (Paper 79-61.3).

Farley, Donna O. 1978. Enumeration and location of hypersusceptible populations. In *Energy Utilization and Environmental Health,* ed. Richard A. Wadden, pp. 17-28. New York: Wiley.

Feldstein, Martin S. 1964. Net social benefit calculation and the public investment decision. *Oxford Economic Papers* 16:114-131.

———. 1972. Distributional equity and the optimal structure of public prices. *AER* 62:32-36.

———. 1974a. Distributional preferences in public expenditure analysis. In *Redistribution Through Public Choice,* ed. Harold M. Hochman and George E. Peterson, pp. 136–161. New York: Columbia Univ. Press.

———. 1974b. Financing in the evaluation of public expenditure. In *Public Finance and Stabilization Policy,* ed. Warren L. Smith and John M. Culbertson, pp. 13–36. Amsterdam: North-Holland Publishing Co.

Ferrar, Terry A., ed. 1976. *The Urban Costs of Climate Modification.* New York: Wiley.

Ferris, Benjamin G. 1978. Health effects of exposure to low levels of regulated air pollutants: A critical review. *JAPCA* 28:482–497.

Fink, F.W.; Buttner, F.H.; and Boyd, W.K. 1971. *Technical-Economic Evaluation of Air Pollution Corrosion Costs on Metals in the U.S.* Columbus, Ohio: Battelle Memorial Inst. (PB 198 453).

Finklea, J.F.; Nelson, W.C.; Moran, J.B.; Akland, G.G.; Larsen, R.I.; Hammer, D.I.; and Knelson, J.H. 1975. Estimates of the public health benefits and risks attributable to equipping light duty motor vehicles with oxidation catalysts. In *Research and Development Related to Sulfates in the Atmosphere,* pp. 943–1015. U.S. House of Representatives, Committee on Science and Technology, 94th Congress, 1st Session.

Fishelson, Gideon, and Graves, Philip. 1978. Air pollution and morbidity: SO_2 damages. *JAPCA* 28:785–789.

Fisher, Anthony C. 1973. Environmental externalities and the Arrow-Lind public investment theorem. *AER* 63:722–725.

Freeman, A. Myrick III. 1972. The distribution of environmental quality. In *Environmental Quality Analysis,* ed. Allen V. Kneese and Blair T. Bower, pp. 243–278. Baltimore: Johns Hopkins Univ. Press.

———. 1979a. *The Benefits of Air and Water Pollution Control: A Review and Synthesis of Recent Estimates.* U.S. Council on Environmental Quality.

———. 1979b. *The Benefits of Environmental Improvement: Theory and Practice.* Baltimore: Johns Hopkins Univ. Press.

Friedman, Milton, and Savage, L.J. 1948. The utility analysis of choices involving risk. *JPE* 56:279–304.

George, P.S., and King, G.A. 1971. *Consumer Demand for Food Commodities in the U.S. With Projections for 1980.* Berkeley, Calif.: Univ. of California, Div. of Agricultural Science (Giannini Foundation Monograph no. 26)

Gerhard, Jon, and Haynie, Fred H. 1974. *Air Pollution Effects on Catastrophic Failure of Metals.* U.S. EPA (EPA 650/3-74-009).

Gillette, Donald G. 1975. Sulfur dioxide and material damage. *JAPCA* 25:1238–1243.

References

———. 1977. Ambient oxidant exposure and health costs in the United States—1973. *JAPCA* 27:329-331.

Goldstein, Inge F., and Landovitz, Leon. 1977. Analysis of air pollution patterns in New York City: I. Can one station represent the large metropolitan area?, *Atmospheric Environment* 11:47-52.

Gregor, John J. 1977. *Intra-Urban Mortality and Air Quality: An Economic Analysis of the Costs of Pollution Induced Mortality.* U.S. EPA (EPA 600/5-77-009).

Guderian, Robert. 1977. *Air Pollution: Phytotoxicity of Gases and its Significance in Air Pollution Control,* trans. C. Jeffrey Brandt. Ecological Studies, vol. 22. New York: Springer Verlag.

Guthrie, Kenneth M. 1969. Capital cost estimating. *Chemical Engineering* 76(6):114-142.

———. 1970. Capital and operating costs for 54 chemical processes. *Chemical Engineering* 77(13):140-156.

Halvorsen, Robert, and Pollakowski, Henry O. 1981. Choice of functional form for hedonic price equations. *J. Urban Economics* 9(3).

Harberger, Arnold C. 1969. On measuring the social opportunity cost of public funds. In *The Discount Rate in Public Investment Evaluation,* Western Agricultural Economics Research Council, Report no. 17, pp. 1-24.

———. 1971. Three basic postulates for applied welfare economics: An interpretive essay. *J. Economic Literature* 9:785-797.

———. 1978. On the use of distributional weights in social cost-benefit analysis. *JPE* 86:S87-S120.

Harrison, David, Jr. 1975. *Who Pays for Clean Air?* Cambridge, Mass.: Ballinger.

Harrison, David, Jr., and Rubinfeld, Daniel L. 1978a. Hedonic housing prices and the demand for clean air. *JEEM* 5:81-102.

———. 1978b. The distribution of benefits for improvements in urban air quality. *JEEM* 5:313-332.

Haynie, Fred H. 1980. Theoretical air pollution and climate effects on materials confirmed by zinc corrosion data. In *Durability of Building Materials and Components,* pp. 157-175. Philadelphia: American Society for Testing and Materials.

Haynie, Fred H.; Spence, James W.; and Upham, James B. 1976. *Effects of Gaseous Pollutants on Materials—A Chamber Study.* U.S. EPA (EPA 600/3-76-015).

Haynie, Fred, H., and Upham, James B. 1970. Effect of atmospheric sulfur dioxide on the corrosion of zinc. *Materials Protection and Performance* 9(8):35-40.

———. 1974. Correlation between corrosion behavior of steel and atmospheric pollution data. In *Corrosion in Natural Environments,* pp. 33–51. Phildelphia: American Society for Testing and Materials.

Heagle, Allen S.; Philbeck, R.B.; and Knott, W.M. 1979. Thresholds for injury, growth, and yield loss caused by ozone on field corn hybrids. *Phytopathology* 69:21–26.

Heagle, Allen S.; Philbeck, R.B.; and Letchworth, Michael B. 1979. Injury and yield responses of spinach cultivars to chronic doses of ozone in open-field chambers. *J. Environmental Quality* 8:368–373.

Heagle, Allen S.; Riordan, A.J.; and Heck, Walter W. 1979. Field methods to assess the impact of air pollutants on crop yields. Paper presented to the Annual Meeting of the Air Pollution Control Assn. (Paper 79-46.6).

Heagle, Allen S.; Spencer, Suzanne; and Letchworth, Michael B. 1979. Yield response of winter wheat to chronic doses of ozone. *Canadian J. Botany* 57:1999–2005.

[Heck, Walter W.] 1977. Plants and microorganisms. In *Ozone and Other Photochemical Oxidants,* pp. 437–585. Washington, D.C.: National Academy of Sciences.

Henderson, James M., and Quandt, Richard E. 1971. *Microeconomic Theory: A Mathematical Approach.* New York: McGraw-Hill Book Co.

Herman, Stewart W. 1977. *The Health Costs of Air Pollution.* New York: American Lung Association.

Hershaft, A.; Morton, J.; and Shea, G. 1976. *Critical Review of Air Pollution Dose-Effect Functions.* U.S. Council on Environmental Quality (PB 251 519).

Hicks, John R. 1939. The foundations of welfare economics. *Economic J.* 49:696–712.

Hill, G.R., Jr., and Thomas, M.C. 1933. Influence of leaf destruction by sulfur dioxide and by clipping on yield of alfalfa. *Plant Physiology* 8:223–245.

Hirshleifer, Jack; Milliman, Jerome W.; and DeHaven, James C. 1960. *Water Supply: Economics, Technology, and Policy.* Chicago: Univ. of Chicago Press.

Hirshleifer, Jack, and Riley, John G. 1979. The analytics of uncertainty and information—An expository survey. *J. Economic Literature* 17:1375–1421.

Holzworth, George C. 1972. *Mixing Heights, Wind Speeds, and Potential for Urban Air Pollution Throughout the Contiguous United States.* U.S. EPA (AP-101).

House, Peter W. 1977. *Trading Off Environment, Economics, and Energy.* Lexington, Mass.: Lexington Books, D.C. Heath and Co.

References

Huber, Joan, ed. 1978. *Marketing Guide to the Paint Industry,* 5th ed. Fairfield, N.J.: Charles H. Kline Co.

Hughes, James P., ed. 1977. Understanding occupational exposure to fluoride. *J. Occupational Medicine* 19:11-87.

Industrial Gas Cleaning Inst. 1972. *Air Pollution Control Technology and Costs in Nine Selected Areas.* U.S. EPA (APTD 1555).

―――. 1973. *Air Pollution Control Technology and Costs in Seven Selected Areas.* U.S. EPA (EPA 450/3-73-010).

ITT Electro-Physics Laboratories. 1971. *A Survey and Economic Assessment of the Effect of Air Pollutants on Electrical Components.* U.S. EPA (APTD-0797).

James, Estelle. 1975. A note on uncertainty and the evaluation of public investment decisions. *AER* 65:200-205.

Johnston, John. 1972. *Econometric Methods,* 2d ed. New York: McGraw Hill.

Jones-Lee, M.W. 1976. *The Value of Life: An Economic Analysis.* Chicago: Univ. of Chicago Press.

Jonsson, E.; Deane, M.; and Sanders, G. 1975. Community reactions to odors from pulp mills. *Environmental Research* 10:249-270.

Kahneman, Daniel, and Tversky, Amos. 1979. Prospect theory: An analysis of decision under risk. *Econometrica* 47:263-292.

Kaldor, Nicholas. 1939. Welfare propositions in economics and interpersonal comparisons of utility. *Economic J.* 49:549-552.

Kaplan, Robert M.; Bush, J.W.; and Berry, Charles C. 1976. Health status: Types of validity and the index of well-being. *Health Services Research* 11:478-507.

Kay, John A. 1972. Social discount rates. *J. Public Economics* 1:359-378.

Kennedy, A.S.; Reisenweber, R.L.; Croke, K.G.; and Snider, M.A. 1973. *An Economic Comparison of Point-Source Controls and Emission Density Zoning for Air Quality Management.* Air Pollution/Land Use Planning Project, vol. 3. U.S. EPA (EPA 450/3-74-028c).

Keyani, Barbara I., and Putnam, Evelyn S. 1977. *Transportation System Management: State of the Art.* U.S. Department of Transportation, Urban Mass Transit Administration (PB 266 953).

Khanna, S.B. 1976. *Handbook for UNAMAP.* Wilmington, Mass.: Walden/Abcor.

Kincannon, Benjamin F., and Castaline, Alan H. 1978. *Information Document on Automobile Emissions Inspection and Maintenance Programs.* U.S. EPA (EPA 400/2-78-001).

Kneese, Allen V. 1968. Comment [on McKean]. In *Problems in Public Expenditure Analysis,* ed. Samuel B. Chase, Jr., pp. 65-71. Washington, D.C.: Brookings Inst.

Kneese, Allen V. and Schultze, Charles L. 1975. *Pollution, Prices, and Public Policy.* Washington, D.C.: Brookings Inst.

Lacasse, Norman, and Treshow, Michael, eds. 1976. *Diagnosing Vegetation Injury Caused by Air Pollution.* Washington, D.C.: U.S. Government Printing Office (Su Doc EP 1.8:V 52).

Larson, Ralph I., and Heck, Walter W. 1976. An air quality data analysis system for interrelating effects, standards, and needed source reductions. Part 3. Vegetation injury. *JAPCA* 26:325-333.

Latimer, D.A., and Samuelsen, G.S. 1978. Visual impact of plumes from power plants: A theoretical model. *Atmospheric Environment* 12: 1455-1465.

Lave, Lester B., and Seskin, Eugene P. 1977. *Air Pollution and Human Health.* Baltimore: Johns Hopkins Univ. Press.

Leaderer, B.P.; Berman, M.D.; and Stolwijk, J.A.J. 1977. Adverse health impact of ambient sulfates, SO_2 and TSP. In *Proceedings of the Fourth International Clean Air Congress.* Tokyo: Japan Union of Air Pollution Prevention Assn.

Leavitt, Jack M.; Leckenby, Henry F.; Blackwell, John P.; and Montgomery, Thomas L. 1974. Cost analysis for development and implementation of a meteorologically scheduled SO_2 emission limitation program for use by power plants in meeting ambient air quality SO_2 standards. *Environmental Letters* 7:135-144.

Lee, R.S., Jr.; Caldwell J.S.; and Morgan, G.B. 1972. The evaluation of methods for measuring suspended particles in air. *Atmospheric Environment* 6:593-622.

Leung, Steve; Goldstein, Elliot; and Dalkey, Norman. 1978. *Human Health Damages from Mobile Source Air Pollution: A Delphi Study.* U.S. EPA (EPA 600/5-78-016a,b).

Leung, Steve; Reed, Walfred; Cauchois, Scott; and Hewitt, Richard. 1978. *Methodologies for Valuation of Agricultural Crop Yield Changes: A Review.* U.S. EPA (EPA 600/5-78-018).

Lindgren, LeRoy H. 1978. *Cost Estimations for Emission Control Related Components/Systems and Cost Methodology Description.* U.S. EPA (EPA 460/3-78-002).

Linnerooth, JoAnne. 1979. The value of human life: A review of the models. *Economic Inquiry* 17:52-74.

Lipfert, Frederick W. 1980. Sulfur oxides, particulates, and human mortality: Synopsis of statistical correlations. *JAPCA* 30:366-371.

Liu, Ben-Chieh, and Yu, Eden S. 1976. *Physical and Economic Damage Functions for Air Pollutants by Receptor.* U.S. EPA (EPA 600/5-76-011).

Loehman, E.T.; Berg, S.V.; Arroyo, A.A.; Hedinger, R.M.; Schwartz, J.M.; Shaw, M.E.; Fahien, R.W.; De, V.H.; Fishe, R.P.; Rio, D.E.;

Rossley, W.F.; and Green, A.E.S. 1979. Distributional analysis of regional benefits and costs of air quality control. *JEEM* 6:222-243.

Luce, R. Duncan, and Raiffa, Howard. 1957. *Games and Decisions: Introduction and Critical Survey.* New York: Wiley.

McGlammery, G.G.; Torstrick, R.L.; Broadfoot, W.J.; Simpson, J.P.; Henson, L.J.; Tomlinson, S.V.; and Young, J.F. 1975. *Detailed Cost Estimates for Advanced Effluent Desulfurization Processes.* U.S. EPA (EPA 600/2-75-006).

McIlvaine, Robert W., and Ardell, Marilyn. 1978. *Research and Development and Cost Projections for Air Pollution Control Equipment.* U.S. EPA (EPA 600/7-78-092).

McKean, Roland N. 1958. *Efficiency in Government Through Systems Analysis.* New York: Wiley.

———. 1968. The use of shadow prices. In *Problems in Public Expenditure Analysis.* ed. Samuel B. Chase, Jr., pp. 33-65. Washington, D.C.: Brookings Inst.

Marder, Sidney M. 1977. *Capital and Operating Cost Equations for Flue Gas Desulfurization Devices.* Chicago: Illinois Inst. for Environmental Quality.

Marglin, Stephen A. 1963a. The social rate of discount and the optimal rate of investment. *QJE* 77:95-111.

———. 1963b. The opportunity costs of public investment. *QJE* 77:274-289.

Margolis, Julius. 1968. Comment [on McKean]. In *Problems in Public Expenditure Analysis,* ed. Samuel B. Chase, Jr., pp. 71-77. Washington, D.C.: Brookings Inst.

Mendelsohn, Robert. 1980. An economic analysis of air pollution from coal-fired power plants. *JEEM* 7:30-43.

———. 1981. The choice of discount rates for public projects. *AER* 71(1).

Mendelsohn, Robert, and Orcutt, Guy. 1977. Pollution dose response curves: A microanalytic study. New Haven, Conn.: Yale University, Inst. for Social and Policy Studies.

———. 1979. An empirical analysis of air pollution dose response curves. *JEEM* 6:85-106.

Mera, Koichi. 1969. Experimental determination of relative marginal utilities. *QJE* 83:464-477.

Merrett, A.J., and Sykes, Allen. 1973. *The Finance and Analysis of Capital Projects,* 2d ed. New York: Wiley.

Michael Baker, Jr., Inc. 1980. *FGD Sludge Disposal Manual,* 2d ed. Palo Alto, Calif.: Electric Power Research Inst. (EPRI FP-977).

Michael, Robert T., and Becker, Gary S. 1973. On the new theory of consumer behavior. *Swedish J. Economics* 75:378-396.

Michaelson, Irving, and Tourin, Boris. 1969. Methodology of determining household costs of air pollution. Paper presented to the Annual Meeting of the Air Pollution Control Assn. (Paper 69-105).

Middleton, W.E.K. 1952. *Vision Through the Atmosphere.* Toronto: Univ. of Toronto Press.

Miedema, A.K.; Decker, C.E.; Smith, F.; and White, J. 1973. *Cost of Monitoring Air Quality in the United States.* U.S. EPA (EPA 450/3-74-029).

Millecan, A.A. 1976. *A Survey and Assessment of Air Pollution Damage to California Vegetation: 1970 Through 1974.* Sacramento, Calif.: California Dept. of Food and Agriculture.

Miller, C. Arthur. 1979. Converting construction costs from one country to another. *Chemical Engineering* 86(14):85-93.

Mishan, E.J. 1967. A proposed normalization procedure for public investment criteria. *Economic J.* 77:777-796.

———. 1977. Economic criteria for intergenerational comparisons. *Zeitschrift für Nationalökonomie* 37:281-306.

Mueller, W.J., and Stickney, P.B. 1970. *A Summary of Economic Assessment of the Effects of Air Pollution on Elastomers.* Columbus, Ohio: Battele Memorial Inst.

National Research Council, Committe on Biologic Effects of Atmospheric Pollutants. 1972. *Particulate Polycyclic Organic Matter.* Washington, D.C.: National Academy of Sciences.

National Research Council, Committee of Medical and Biologic Effects of Environmental Pollutants. 1977. *Arsenic.* Washington, D.C.: National Academy of Sciences.

———. 1978. *Hydrogen Sulfide.* Baltimore: University Park Press.

———. 1979a. *Airborne Particles.* Baltimore: University Park Press.

National Research Council, Advisory Committee on the Biological Effects of Ionizing Radiation. 1979b. *Effects on Populations of Exposure to Low Levels of Ionizing Radiation.* Washington, D.C.: National Academy of Sciences.

Needleman, Lionel. 1976. Valuing other people's lives. *Manchester School of Economic and Social Studies* 44:309-342.

Nelson, Jon P. 1978. Residential choice, hedonic prices, and the demand for urban air quality. *J. Urban Economics* 5:357-369.

Nelson, William C.; Knelson, John H.; and Hasselblad, Victor. 1976. Air pollutant health effects estimation model. In *Proceedings of the Conference on Environmental Modeling and Simulation,* pp. 191-195. U.S. EPA (EPA 600/9-76-016).

Neveril, R.B. 1978. *Capital and Operating Costs of Selected Air Pollution Control Systems,* 2d ed. U.S. EPA (EPA 450/5-80-002).

Neveril, R.B.; Price, J.U.; and Engdahl, K.L. 1978. Capital and operating costs of selected air pollution control systems. *JAPCA* 28:829-836, 963-968, 1069-1072, 1171-1174, 1253-1256.

North, D. Warner, and Merkhofer, M.W. 1975. Analysis of alternative emissions control strategies. In *Air Quality and Stationary Source Emission Control,* chap. 13. U.S. Senate, Committee on Public Works, 94th Congress, 1st Session (Serial 94-4).

O'Connor, J.A.; Parbery, D.G.; and Strauss, W. 1974. The effects of phytotoxic gases on Australian native plant species. I. Acute effects of sulfur dioxide. *Environmental Pollution* 7:7-23.

———. 1975. The effects of phytotoxic gases on Australian native plant species. II. Acute injury due to ozone. *Environmental Pollution* 9:181-192.

Olson, Craig A. 1981. An analysis of wage differentials received by workers on dangerous jobs. *J. Human Resources* 16:167-185.

Ormrod, D.P. 1978. *Pollution in Horticulture.* Fundamental Aspects of Pollution Control and Environmental Science, vol. 4. New York: Elsevier Scientific.

Oshima, R.J.; Braegelmann, P.K.; Baldwin, D.W.; Van Way, V.; and Taylor, O.C. 1977. Reduction of tomato fruit size and yield by ozone. *J. American Society for Horticultural Science.* 102:289-293.

Oshima, R.J.; Poe, M.P.; Braegelmann, P.K.; Baldwin, D.W.; and Van Way, V. 1976. Ozone damage-crop loss function for alfalfa: A standardized method for assessing crop losses from air pollutants. *JAPCA* 26:861-865.

Page, Talbot. 1977. Equitable use of the resource base. *Environment and Planning A* 9:15-22.

Palmini, Dennis, and Rossi, Daniel. 1980. What price air quality? The cost of New Jersey's inspection/maintenance program. *JAPCA* 30:1081-1088.

Park, William R. 1973. *Cost Engineering Analysis.* New York: Wiley.

PEDCo Environmental, Inc. 1978. *Particulate and Sulfur Dioxide Emission Control Costs for Large Coal-Fired Boilers.* U.S. EPA (EPA 450/3-78-007).

Peters, Max S., and Timmerhaus, Klaus D. 1979. *Plant Design and Economics for Chemical Engineers,* 3d ed. New York: McGraw Hill.

Peterson, J.E., and Stewart, R.D. 1975. Predicting the carboxyhemoglobin levels resulting from carbon monoxide exposures. *J. Applied Physiology* 39:633-638.

Phlips, Louis, 1974. *Applied Consumption Analysis.* Amsterdam: North-Holland Publishing Co.

Pikulik, Arkadie, and Diaz, Hector E. 1977. Cost estimating for major process equipment. *Chemical Engineering* 84(21):65-122.

Ponder, Thomas C.; Yerino, Lario V.; Katari, Vishnu; Shah, Yatendra; and Devitt, Timothy W. 1976. *Simplified Procedures for Estimating Flue Gas Desulfurization System Costs.* U.S. EPA (EPA 600/2-76-150).

Rawls, John. 1971. *A Theory of Justice.* Cambridge, Mass.: Harvard Univ. Press.
Ricci, Paolo F. and Wyzga, Ronald E. 1979. Review of epidemiological studies of air pollution: Statistical emphasis. Paper presented to the Meeting on Benefit Methodology, U.S. National Commission on Air Quality.
———. 1981. A statistical review of cross-sectional studies of ambient air pollution and mortality. *Environment International* (in press).
Ridker, Ronald G. 1967. *Economic Costs of Air Pollution.* New York: Praeger.
Roeck, D.R., and Dennis, R. 1979. *Technology Assessment Report for Industrial Boiler Applications: Particulate Collection.* U.S. EPA (EPA 600/7-79-178b).
Rose-Ackerman, Susan. 1973. Effluent charges: A critique. *Canadian J. of Economics* 6:512-528.
Rowe, Robert D.; d'Arge, Ralph C.; and Brookshire, David S. 1980. An experiment on the economic value of visibility. *JEEM* 7:1-19.
Rubinfeld, Daniel L. 1978. Market approaches to the measurement of the benefits of air pollution abatement. In *Approaches to Controlling Air Pollution,* ed. Ann F. Friedlaender, pp. 240-273. Cambridge, Mass.: MIT Press.
Ruby, Michael G. 1978. An application of benefit-cost analysis: The ASARCO-Tacoma copper smelter. Paper presented to the Annual Meeting of the PNWIS—Air Pollution Control Assn. (Paper PNWIS 78-12).
Rutledge, Gary L. 1979. Pollution abatement and control expenditures in constant and current dollars, 1972-77. *Survey of Current Business* 59(2):13-20.
Rutledge, Gary L., and O'Connor, Betsy. 1979. Capital expenditures by business for pollution abatement, 1977, 1978, and planned 1979. *Survey of Current Business* 59(6):20-22.
Salmon, Richard L. 1970. *Systems Analysis of the Effects of Air Pollution on Materials.* Kansas City: Midwest Research Inst. (PB 209 192).
Samuelson, Paul A. 1955. Diagrammatic exposition of a theory of public expenditure. *Review of Economics and Statistics* 37:350-356.
———. 1956. Social indifference curves. *QJE* 70:1-22.
Sandmo, Agnar, and Dreze, Jacques H. 1971. Discount rates for public investment in closed and open economies. *Economica* 38:395-412.
Schimmel, Herbert, and Murawski, Thaddeus J. 1976. The relation of air pollution to mortality. *J. Occupational Medicine* 18:317-333.
Schroeder, W.H. 1975. *Air Pollution Aspects of Odorous Substances.* Canada, Department of Fisheries and Environment (EPS 3-AP-75-1).

Schwing, Richard C. 1979. Longevity benefits and the cost of reducing various risks. *Technogical Forecasting and Social Change* 13:333-345.
Schwing, Richard C., and McDonald, Gary C. 1976. Measures of association of some air pollutants, natural ionizing radiation and cigarette smoking with mortality rates. *The Science of the Total Environment* 5:139-169.
Schwing, Richard C.; Southworth, Bradford W.; von Buseck, Calvin R.; and Jackson, Clement J. 1980. Benefit-cost analysis of automotive emission reductions. *JEEM* 7:44-64.
Scitovsky, Tibor. 1941. A note on welfare propositions in economics. *Review of Economic Studies* 9:77-88.
Scott, R. 1978. Working capital and its estimation for project evaluation. *Engineering and Process Economics* 3:104-114.
Sen, Amartya K. 1967. Isolation, assurance, and the social rate of discount. *QJE* 81:112-124.
Senew, Michael; Donaldson, Thomas; Conley, Lester; Cirillo, Richard; Seymour, Dorothea; and Smith, Donald. 1978. Estimating the cost to states to implement federal air pollution control strategies. Paper presented to the Annual Meeting of the Air Pollution Control Assn. (Paper 78-3.2)
Shy, Carl M. 1979. On using epidemiologic evidence of air pollution effects. Paper presented to the Meeting on Benefit Methodology, U.S. National Commission on Air Quality.
Shy, Carl M.; Goldsmith, John R.; Hackney, Jack D.; Lebowitz, Michael D.; and Menzel, Daniel B. 1978. *Health Effects of Air Pollution.* New York: American Lung Association.
Skelly, John M.; Krupa, Sagar V.; and Chevone, Boris I. 1979. Field surveys. In *Methodology for the Assessment of Air Pollution Effects on Vegetation,* ed. W.W. Heck, S.V. Krupa, and S.N. Linzon, chap. 12. Pittsburgh: Air Pollution Control Assn.
Slovic, Paul; Lichtenstein, Sarah; and Fischhoff, Baruch. 1979. Images of disaster: Perception and acceptance of risk from nuclear power. In *Energy Risk Management,* ed. G.T. Goodman and W.D. Rowe, pp. 223-245. London: Academic Press.
Smith, Barton A. 1978. Measuring the value of urban amenities. *J. Urban Economics* 5:270-287.
Smith, Robert S. 1979. Compensating wage differentials and public policy: A review. *Industrial and Labor Relations Review* 32:339-352.
Spence, James W., and Haynie, Fred H. 1972. *Paint Technology and Air Pollution: A Survey and Economic Assessment.* U.S. EPA (AP-103).
Spengler, John D.; Colombe, Steven D.; Evans, John S.; Dawson, Stanley V.; Dockery, Douglas W.; and Thibodeau, Lawrence A. 1979. Quanti-

tative assessment of the association of particulates and sulfates with total mortality: Cross-sectional studies. In *The Direct Use of Coal: Prospects and Problems of Production and Combustion,* app. IX, chap. 5. U.S. Congress, Office of Technology Assessment (PB 80-184526).

Stern, Arthur C., ed. 1976. *Air Pollutants, Their Transformation and Transport.* Air Pollution, 3d ed. vol. 1. New York: Academic Press.

———. 1977a. *The Effects of Air Pollution.* Air Pollution, 3d ed., vol. 2. New York: Academic Press.

———. 1977b. *Engineering Control of Air Pollution.* Air Pollution, 3d ed., vol. 4. New York: Academic Press.

Stevens, R.W. 1947. Equipment cost indexes for process industries. *Chemical Engineering* 54(11):124-126.

Systems Applications, Inc. 1975. *An Examination of the Accuracy and Adequacy of Air Quality Models and Monitoring Data for Use in Assessing the Impact of EPA Significant Deterioration Regulations on Energy Development.* Washington, D.C.: American Petroleum Institute (EF75-58R).

Thaler, Richard, and Rosen, Sherwin. 1976. The value of saving a life: Evidence from the labor market. In *Household Production and Consumption.* ed. Nester E. Terleckyj, pp. 265-298. Studies in Income and Wealth, vol. 40. New York: Columbia Univ. Press.

Theil, Henri. 1975. *Theory and Measurement of Consumer Demand.* Amsterdam: North-Holland Publishing Co.

Thibodeau, Lawrence A.; Reed, R.B.; Bishop, Yvonne M.M.; and Kammerman, L.A. 1980. Air Pollution and Human Health: A review and reanalysis. *Environmental Health Perspectives* 34:165-183.

Tietenberg, Thomas H. 1978. Spatially differentiated air pollutant emission charges: An economic and legal analysis. *Land Economics* 54: 265-277.

Tingey, David T., and Reinert, Richard A. 1975. The effect of ozone and sulfur dioxide singly and in combination on plant growth. *Environmental Pollution* 9:117-125.

Torstrick, R.C.; Henson, L.J.; and Tomlinson, S.U. 1978. Economic evaluation techniques, results and computer modeling for flue gas desulfurization. In *Proceedings: Symposium on Flue Gas Desulfurization,* vol. 1, pp. 118-168. U.S. EPA (EPA 600/7-78-058a).

Turner, D. Bruce. 1970. *Workbook of Atmospheric Dispersion Estimates.* U.S. EPA (AP-26).

———. 1979. Atmospheric modeling: A critical review. *JAPCA* 29: 502-519.

Tversky, Amos, and Kahneman, Daniel. 1981. The framing of decisions and the psychology of choice. *Science* 211:453-458.

References

Uhl, Vincent W. 1979. *A Standard Procedure for Cost Analysis of Pollution Control Operations,* vols. 1 and 2. U.S. EPA (EPA 600/8-79-018a,b).

U.S. Council on Environmental Quality. 1978. *Environmental Quality: The Ninth Annual Report of the Council on Environmental Quality.*

U.S. Department of Commerce, Bureau of the Census. Annual, 1975-1979. *Selected Industrial Air Pollution Control Equipment* (MA-35J).

U.S. Department of the Treasury, Internal Revenue Service. 1979. *Tax Information on Depreciation* (Pub. 534).

U.S. Environmental Protection Agency. 1969. *Control Techniques for Particulate Pollutants* (AP-51).

———. 1972. *Air Quality Data for Sulfur Dioxide: 1969, 1970, 1971* (APTD 1354).

———. 1974a. *Background Information for New Source Performance Standards: Primary Copper, Zinc, and Lead Smelters* (EPA 450/2-74-002a).

———. 1974b. *Health Consequences of Sulfur Oxides: A Report from CHESS, 1970-1971* (EPA 650/1-74-004).

———. 1976. *Monitoring and Air Quality Trends Report, 1974* (EPA 450/1-76-001).

———. 1977a. *Air Quality Criteria for Lead* (EPA 600/8-77-017).

———. 1977b. *Control Techniques for Lead Air Emissions* (EPA 450/2-77-012).

———. 1977c. *Guidelines on Procedures for Constructing Air Pollution Isopleth Profiles and Population Exposure Analysis* (EPA 450/2-77-024a).

———. 1977d. *Industrial Process Profiles for Environmental Use,* 26 vols (EPA 600/2-77-023).

———. 1977e. *Uses, Limitations, and Technical Basis for Procedures for Quantifying Relationships Between Photochemical Oxidants and Precursors* (EPA 450/2-77-021a and Supplement).

———. 1978a. *Air Quality Criteria for Ozone.* (EPA 600/8-78-004).

———. 1978b. *Control Techniques for Nitrogen Oxides Emissions From Stationary Sources,* 2d ed. (EPA 450/1-78-001).

———. 1978c. *Control Techniques for Volatile Organic Emissions From Stationary Sources* (EPA 450/2-78-022).

———. 1978d. *Electric Utility Steam Generating Plants: Background Information for Proposed Emission Standards* (EPA 450/2-78-005a, -006a, -007a, -007a-1).

———. 1978e. *Mobile Source Emission Factors* (EPA 400/9-78-006).

———. 1978f. *Transit Improvement, Preferential Lane, and Car Pool Programs* (EPA 400/2-78-002a).

———. 1979a. *Air Quality Criteria for Carbon Monoxide* (EPA 600/8-79-022).

---. 1979b. *Control Techniques for Carbon Monoxide Emissions* (EPA 450/3-79-006).

---. 1979c. *The Cost of Clean Air and Water: Report to Congress* (EPA 230/3-79-001).

---. 1979d. *Technology Assessment Report for Industrial Boiler Applications* (EPA 600/7-79-178b-i).

---. 1979e. Financial and economic analysis techniques. In *Guidance for Lowest Achievable Emission Rates from 18 Major Stationary Sources of Particulate, Nitrogen Oxides, Sulfur Dioxide, or Volatile Organic Compounds,* section 5 (EPA 450/3-79-024).

---. 1980a. *Air Quality Criteria for Nitrogen Dioxide.*

---. 1980b. *Compilation of Air Pollutant Emission Factors (Including Supplements 1-11)* (AP-42).

---. 1980c. *Control Techniques for Sulfur Dioxide Emissions from Stationary Sources,* 2d ed. (EPA 600/8-78-004).

---. 1980d. *Proposed Guidelines for Determining Best Available Retrofit Technology for Coal-Fired Power Plants and Other Major Stationary Sources* (EPA 450/3-80-009d).

---. 1981. *Air Quality Criteria for Particulate Matter and Sulfur Oxides.*

U.S. House of Representatives, Committee on Interstate and Foreign Commerce, Subcommittee on Oversight and Investigations. 1976a. *Federal Regulation and Regulatory Reform,* chap. 15. 94th Congress, 2nd Session (Subcommittee Print).

---. Committee on Science and Technology. 1976b. *Community Health and Environmental Surveillance System (CHESS): An Investigative Report.* 94th Congress, 2nd Session.

U.S. Interagency Task Force on Motor Vehicle Goals Beyond 1980. 1976. *Air Quality, Noise and Health.* U.S. Department of Transportation, Office of the Secretary of Transportation.

U.S. National Institute for Occupational Safety and Health. 1974. *Criteria for a Recommended Standard: Occupational Exposure to Benzene* (NIOSH/74-137).

---. 1976. *Criteria for a Recommended Standard: Occupational Exposure to Cadmium* (NIOSH 76-192).

---.1978. *Criteria for a Recommended Standard: Occupational Exposure During the Manufacture and Formulation of Pesticides* (NIOSH 78-174).

U.S. Water Resources Council. 1974. *1972 OBERS Projections: Economic Activity in the United States.*

Viscusi, W. Kip. 1978. Labor market valuations of life and limb: Empirical evidence and policy implications. *Public Policy* 26:359-386.

von Neumann, John, and Morgenstern, Oskar. 1947. *Theory of Games and Economic Behavior,* 2d ed. Princeton, N.J.: Princeton Univ. Press.

References

Waggoner, Alan P., and Weiss, Ray E. 1980. Comparisons of fine particle mass concentrations and light scattering extinction in ambient aerosol. *Atmospheric Environment* 14:623-626.

Walsh, Phillip J.; Killough, George G.; and Rohwer, Paul S. 1978. Composite hazard index for assessing limiting exposures to environmental pollutants: Formulation and derivation. *Environmental Science and Technology* 12:799-802.

Watson, William D., and Jaksch, John A. 1978. Household cleaning costs and air pollution. Paper presented to the Annual Meeting of the Air Pollution Control Assn. (Paper 78-52.3).

Weaver, James B., and Bauman, H. Carl. 1973. Cost and profitability estimation. In *Chemical Engineering Handbook,* 5th ed., ed. Robert H. Perry and Cecil H. Chilton, section 25. New York: McGraw Hill.

Weisbrod, Burton A. 1968. Income redistribution effects and benefit-cost analysis. In *Problems in Public Expenditure Analysis,* ed. Samuel B. Chase, Jr., pp. 177-222. Washington, D.C.: Brookings Inst.

Weiss, Ray E.; Waggoner, Alan P.; Charlson, Robert J.; Thorsell, David L.; Hall, J.S.; and Riley, L.A. 1979. Studies of the optical, physical and chemical properties of light absorbing aerosols. In *Proceedings: Carbonaceous Particles in the Atmosphere,* ed. T. Novakov. Berkeley, Calif.: Lawrence Berkeley Laboratory (CONF 7803101).

Weitzman, Martin L. 1974. Prices vs. quantities. *Review of Economic Studies* 41:477-491.

Wesolowski, Wayne E. 1977. *Initial Development and Operating Costs for Supplementary Control Systems.* Chicago: Illinois Inst. for Environmental Quality.

Westman, Walter E. 1977. How much are nature's services worth? *Science* 197:960-964.

Williams, Alan. 1972. Cost-benefit analysis: Bastard science? And/or insidious poison in the body politick? *J. Public Economics* 1:199-225.

Williams, Roger, Jr. 1947. Six-tenths factor aids in approximating costs. *Chemical Engineering* 54(December):124-125.

Willig, Robert D. 1976. Consumer's surplus without apology. *AER* 66:589-597.

Willig, Robert D., and Bailey, Elizabeth E. 1981. Income distributional concerns in regulatory policy-making. In *Studies in Public Regulation,* ed. Gary Fromm. Cambridge, Mass.: MIT Press.

Wilson, Richard. 1978. Risks caused by low levels of pollution. *Yale J. Biology and Medicine* 51:37-51.

Woods, Donald R. 1975. *Financial Decisions in the Process Industries.* Englewood Cliffs, N.J.: Prentice-Hall.

Woods, Donald R.; Anderson, Susan J.; and Norman, Suzanne L. 1976. Evaluation of capital cost data: Heat exchangers. *Canadian J. Chemical Engineering* 54:469-488.

———. 1978. Evaluation of capital cost data: Gas moving equipment. *Canadian J. Chemical Engineering* 56:413–435.

Wyzga, Ronald E. 1978. The effect of air pollution upon mortality: A consideration of distributed lag models. *J. American Statistical Association* 73:463–472.

Yen, Yen-Chen. 1972. Estimating plant costs in developing countries. *Chemical Engineering* 79(15):89–92.

Yocom, John E., and Grappone, Nicola. 1976. *Effects of Power Plant Emissions on Materials.* Palo Alto, Calif.: Electric Power Research Inst. (EPRI EC-139).

Zeckhauser, Richard, and Shepard, Donald. 1976. Where now for saving lives? *Law and Contemporary Problems* 40:5–45.

Zielhuis, R.L., ed. 1977. *Public Health Risks of Exposure to Asbestos.* Oxford: Pergamon Press.

Index

Absorber, 190, 224–225
Acetates, 150
Acid rain, 140, 205, 214
Acton, Jan Paul, 126
Adams, Richard M., Narongsakdi Thanavibulchai, and Thomas D. Crocker, 143
Adsorber, 182, 224
Aesthetic benefits, 2, 142; estimating, 155–167
Afterburner, 185, 224–226
Aggregate consumers' and producers' surplus, 80–83
Aggregation over individuals, 29–36; with explicit distributional weights, 32–35; without explicit distributional weights, 30–32
Aggregation over time, 37–62
Agricultural data, 104, 133–143, 201–202, 207, 214
Air pollutants: sources, measurement, and effects, 197–214
Air-pollution control, optimal degree of, 2–5
Air-pollution-control equipment, 169, 215–227; cost functions, 181–189; life time, 176; purchases, 216
Air Pollution Control Strategy Resource Estimator, 192
Air-pollution effects, 197–214; quantifying, 99–110
Air-quality-monitoring stations, 102–103, 120, 121, 192–193
Alfalfa, 135, 138, 202, 207, 214
Allais, Maurice, 69
Aluminum, 146, 201
Ambient-air-quality standards, 99, 102–103, 106–107, 121
American Lung Association, 197
Anderson, David, 171
Anemia, 203
Angina pectoris, 124
Animal and ecosystems dose-response functions, 140
Approximation formulas (CPS), 87–94
Argonne National Laboratory, 191, 192
Arnold, Thomas H., and Cecil H. Chilton, 176
Arrow, Kenneth J., 26, 62
Arrow, Kenneth J., and Robert C. Lind, 74
Arsenic, 197–198

Asbestos, 198–199
Asthma, 123, 213
Atmospheric stability, 100
Automobile-emission control systems, 194–195
Auxiliary equipment and ducting, 226
Azzi, Corry F., and James C. Cox, 27

Babcock, Lyndon R., and Niren L. Nagda, 111
Baghouse, 182–183, 185, 220–221
Bailey, Martin J., 26, 126
Baram, Michael S., 8
Bator, Francis M., 26
Baumol, William J., 26
Baumol, William J., and Wallace E. Oates, 10
Bees, 140, 202
Beloin, Norman J., 151
Beloin, Norman J., and Fred H. Haynie, 156, 158
Benarie, M., 102, 103
Benedict, Harris M., and John A. Jaksch, 133
Benedict, Harris M., Clarence J. Miller, and Jean S. Smith, 120, 138–139, 142
Benefit-cost analysis: elements of, 5–8; criticisms of, 8–9
Benefit-cost-ratio decision rules, 46–48, 52
Benefits and costs: aggregation over individuals, 29–36; aggregation over time, 37–62; marginal, 2–3; uncertain, evaluation of, 63–75
Benzene, 199
Benzo(a)pyrene, 211
Bergson, Abram, 26
Bidding studies, 125–126, 161–163
Blecker, Herbert G., and Theodore W. Cadman, 171
Blecker, Herbert G., and Thomas M. Nichols, 190
Bloom, S.G., 174–175, 182
Blood, diseases of, 104, 199, 203, 211
Blood concentrations, 112
Boadway, Robin W., 33–35, 36, 62
Bone cancer, 212
Bone damage, 200
Bone marrow, injury to, 211
Booz, Allen, and Hamilton, Inc., 157, 158
Boston, 166
Box model, 100

Bradford, David F., 59, 62
Breast cancer, 211
"British Smoke," 208
Bronchitis, 118, 123, 204, 209, 213
Brookshire, David S., 163, 164
Brown, Charles, 126
Browning, Edgar K., 25
Brysson, Ralph J., 151
Building stone, 145, 151-152
Burchard, John K., 184-186
Burchard-Marder equation, 186-189

C-E Air Preheater, 190
Cadmium, 199-200
California, 140, 152
Campbell, G,G., 150
Canada, 195, 214
Cancer, 105, 198-199, 200, 205, 211, 212
Capital and Operating Costs of Selected Air Pollution Control Systems (Neveril), 189
Capital costs, direct/indirect, 173, 174, 191-194
Capital markets, perfect, 40, 53
Capital recovery factor, 153
Car-pool programs, 195
Carbon monoxide, 112, 123, 124, 200-201
Carboxyhemoglobin, 200-201
Carlson, John A., 60
Caspari, Conrad, Donald MacLaren, and Georgina Hobhouse, 143
Cattle, 140, 142, 202
Census-tract property values, 165-166
Central-city bias, 119, 120
Certainty equivalent, 68
Chapman, Richard A., 191
Chemical Engineering (CE) Plant Index, 176-178
Chicago, 117, 166
Clean Air Act, xv, 106, 197
Clean Air Amendments (1970), xv
Climate, changes in, 142
Coal-characteristics data base, 191
Coase, Ronald, 4-5
Coefficient-of-haze (COH), 117, 130, 208
Cohen, Alan S., 8
Cohen, Alan S., Gideon Fishelson, and John L. Gardner, 7-8
Commodities, priced/unpriced, valuation of, 77-96
Commodity taxes, 84
Compensating variation, 79
Compensating wage premiums, 126-128
Construction-price indexes, 178
Consumer's and producer's surplus, estimates of, 87-94
Consumer's surplus, 78-79, 174
Contingency reserves, 174

Control costs: data for estimating, 181-195; evaluation of, 169-180; financial analysis of, 178-179
Control efficiency, 171
Control equipment cost functions, 181-188, 229-236
Control requirements, defining, 171-172
Control Techniques (EPA), 172
"Conventional"/"nonconventional" distribution of net benefits over time, 49, 62
Cooper, Barbara S., and Dorothy P. Rice, 129
Copper, 146
Corn, 207
Corrosion, 146-150
Cost data, miscellaneous, 191-194
Cost estimates: items included in, 172-178; sources of variability in, 174-176; types of, 169-171
Cost functions, 171; control-equipment, 184; flue-gas-desulfurization, 184-189; order-of-magnitude estimates, 181-189, 229-236
Cost multipliers, 170-171, 184-185, 192
"Cost of Clean Air" project (EPA), 183, 229-236
Cotton fabric, 145, 150, 151
Cran, John, 178
Crittenden, P.D., and D.J. Read, 138
Crocker, Thomas D., 117, 118, 119, 121
Cross-section data, 108
Culyer, A.J., 26
Cyclone collector, 182, 185, 215-218

Damage function, 105, 111
Danielson, John A., 227
Davis, C.R., 138
Davis, Donald D., and Raymond G. Wilhour, 134
Debt-service-coverage ratio, 178
Decision rules, 45-52; benefit-cost ratio, 46-48; comparison, 52; expected monetary value, 63-65; expected utility, 65-69; internal-rate-of-return, 48-50; net-present-value, 45-46; payback-period, 50-52
Definitive or detailed estimate (control costs), 170
Demand and supply curves, 78-83
Demand and supply elasticities: agricultural crops, 143; defined, 88; environmental quality, 165-166
Demandel, R.E., 103
Depreciation allowances, 174, 176
Devitt, Timothy W., 175
Devitt, Timothy W., P. Spaite, and L.

Index   259

Gibbs, 195
Dewees, Donald, 194
Dichotomous sampler, 208
Diet data, 119, 121
Discount rate, 42-43, 51, 52-62
Discounted cash flow, 179
Dispersion models, 100-102
Distribution of benefits and costs, 22-25, 29-36
Dorfman, Nancy S., and Arthur Snow, 35
Dose-response functions, 104-107; animal and ecosystems, 140; health effects, 111-113, 120-121; metals, 146-149; paint, 149-150; public-annoyance, 155; subjective assessment of, 122-124; textiles, 150-151; vegetation, 133-139; visibility, 161; yield, 134-135
Double-alkali-control process, 186, 187, 190, 191
Dravnieks, Andrew, and Hugh J. O'Neill, 155
Dustfall jar, 209
Dynamic efficiency, concept of, 38-41
Dynamic inefficiency: implications of, 54-57; sources of, 52-54

Economic efficiency, defined, 14
Economic-efficiency criterion, 14-17
Economic-welfare criterion, 19-20, 30-31; and potential-compensation criteria compared, 21-23
Ecosystems: damages, 140; valuation, 142
Edmiston, Norman G., and Francis L. Bunyard, 184
Edwards, Clive Arthur, 210
El-Sawy, A., J.G. Leigh, and R.K. Trehan, 175
Elasticity, defined, 88
Electric-power plants, 174-175, 184-189
Electronic equipment, 152
Electrostatic precipitators, 182, 185, 190, 191, 195, 218-220
Emission factors, 171
Emission inventory, 171
Emission standard, 99
Employment data, 103-104
Enforcement costs, 173, 174, 192
Ensor, David S., Leslie E. Sparks, and Michael J. Pilat, 161
Environmental quality (as an unpriced input), 85-87
Epidemiologic studies, 107, 112-121
Equivalent variation, 79
Error, sources of: in macroepidemiologic studies, 119-120; in regression analysis, 108-110
"Evaluation of Capital Cost Data"

(CJCE), 189
Expected monetary value (EMV), 63-65, 70-74
Expected utility, 65-69, 71-73
Extinction coefficient (light), 160
Eye irritation, 123, 207

Fabric dyes, 150, 151, 205
Fabric filters, 182-183, 185, 190, 191, 220-221
Fabricated Equipment Index, 178
Fabrik, A., R. Skarlew, and J. Wilson, 102
Fama, Eugene F., 60
Fans, 226
Farber, Paul S., 191
Farber, Paul S., and C.D. Livengood, 191
Farley, Donna O., 103
Farmington, N.M., 162-163
Feldstein, Martin S., 32-33, 36, 59, 62
Ferrar, Terry A., 142
Ferris, Benjamin G., 197
Field studies, 107, 138, 139-140
Fine-particle concentration, 120, 160-161, 167, 207-210
Fink, F.W., F.H. Buttner, and W.K. Boyd, 147, 154
Finklea, J.F., 7
Fishelson, Gideon, and Philip Graves, 118
Fisher, Anthony C., 74
Flange-to-flange cost, 170
Flue-gas-desulfurization (FGD) cost functions, 184-189
Fluoride, 134, 138, 139, 140, 201-202
Forests, 140
Freeman, A. Myrick, III, 1, 35, 85, 95, 126, 165
Friedman, Milton, and L.J. Savage, 68-69, 75
Fuel-desulfurization unit, 183

Galvanizing, 146-147, 148-149
Gas preconditioning, 226
Gaussian plume model, 100-102
George, P.S., and G.A. King, 143
Gerhard, Jon, and Fred H. Haynie, 147
Gillette, Donald G., 124, 149
Goldstein, Inge F., and Leon Landovitz, 120
"Grass-roots" or "greenfields" site, 170
Great Britain, 209
Gregor, John J., 120, 121
Guderian, Robert, 134
Guthrie, Kenneth M., 171, 190, 192

Halvorsen, Robert, and Henry O. Pollakowski, 110
Handling equipment, solids and liquids, 190

Harberger, Arnold C., 36, 62
Harrison, David, Jr., 7
Harrison, David, Jr., and Daniel L. Rubinfeld, 35, 166
Haynie, Fred H., 148
Haynie, Fred H., James W. Spence, and James B. Upham, 151
Haynie, Fred H., and James B. Upham, 147
Hazardous pollutants, 197
Health benefits, 2; estimating, 111-131
Health damages, valuation of, 124-129, 163
Health-effects dose-response functions, development of, 111-113
Heart disease, 118, 123, 124, 200, 204
Herbicides, 210
Heck, Walter W., 134, 135-137
Henderson, James M., and Richard E. Quandt, 26, 27
Herman, Stewart W., 113
Hershaft, A., J. Morton, and G. Shea, 110
Hicks, John R., 20-21
Hill, G.R., Jr., and M.C. Thomas, 138
Hirshleifer, Jack, Jerome W. Milliman, and James C. Dehaven, 62
Hirshleifer, Jack, and John G. Riley, 69, 75
Hockey-stick function, 105
Holzworth, George C., 139
Horses, 203
House, Peter W., 183
Household-cleaning-operations costs, 156-158
Household production concept, 87, 118
Huber, Joan, 154, 167
Hughes, James P., 202
Human capital, 128-129
Hydrocarbons, 139
Hydrogen sulfide, 147, 202-203

Incinerators, thermal/catalytic, 185, 190, 224-226
Income effects, 162, 165-166
Indifference curves, 38
Inflation, 44, 175-178
Injury, visible (vegetation), 135-138
Insect damage, plant sensitivity to, 140
Insecticides, 210
Installed cost, 170, 172
Interest-during-construction, 174, 186
Interest rate, real/nominal, 44-45, 62, 176
Interest rate earned on savings, 60
Intermittent control system (ICS), 192
Internal-rate-of-return decision rules, 48-50
Investment decision rules, 45-52, 57-62
Involuntary risks, 128
Iron, 146
ITT Electro-Physics Laboratories, 152

Jacobs-Hochheiser method, 109, 204
James, Estelle, 75
Johnston, John, 110
Jones-Lee, M.W., 126
Jonsson, E., M. Deane, and G. Sanders, 155

Kahneman, Daniel, and Amos Tversky, 69-70
Kaldor, Nicholas, 20-21, 23-25
Kaplan, Robert M., J.W. Bush, and Charles C. Berry, 128
Kay, John A., 62
Kennedy, A.S., 184
Keyani, Barbara I., and Evelyn S. Putnam, 195
Khanna, S.B., 102
Kidney cancer, 211
Kidney damage, 200
Kincannon, Benjamin F., and Alan H. Castaline, 194
Kneese, Allen V., 95
Kneese, Allen V., and Charles L. Schultze, 10

Labor and overhead cost, 173, 188
Labor-materials ratios, 171, 176, 178
Labor-supply curve, 79-80
Lacasse, Norman, and Michael Treshow, 134
Land costs, 166, 192
Larsen, Ralph I., and Walter W. Heck, 136-137
Latimer, D.A., and G.S. Samuelson, 161
Lave, Lester B., and Eugene P. Seskin, 116, 117, 120, 121
Law of large numbers, 64
Lead, 146
Lead pollution, 104, 112, 203
Leaderer, B.P., M.D. Berman, and J.A.J. Stolwijk, 118
Leaf injury, 135-138
Leavitt, Jack M., 192
Lee, R.S., Jr., J.S. Caldwell, and G.B. Morgan, 130
Leukemia, 199, 212
Leung, Steve, 143
Leung, Steve, Elliot Goldstein, and Norman Dalkey, 124
Lichens, 140, 214
Life: equipment, 176; mortality studies, 113-117; value of, 124-129
Light scattering and visibility, theory of, 159-161
Lime/limestone control processes, 186, 187, 190, 191
Linear dose response function, 105-106

Index

Lingren, LeRoy H., 194
Linnerooth, JoAnne, 126
Liu, Ben-Chieh, and Eden S. Yu, 116, 121, 139, 157
Loehman, E.T., 8, 35, 125, 161-162
Los Angeles, 134, 141, 163
Loss factors (vegetation), 138-139
Luce, R. Duncan, and Howard Raiffa, 75
Lung cancer, 198-199, 211, 212
Lung diseases, 111-112, 118, 123, 200, 204, 206-207, 209, 211, 213
Lung function, 111-112, 209, 213

M & S Index, 176-178
Macroepidemiologic studies, 113-121; morbidity studies, 117-118; mortality studies, 113-117
Magnesium-oxide control process, 190, 191
Maintenance factor, 188-189
Marder, Sidney M., 186, 187
Marginal benefits and costs, 2-3
Marginal rate of substitution, 38
Marginal rate of transformation, 38
Marginal social significance, 17, 23
Marginal social utility of income, 22-23, 29-30, 31
Marglin, Stephen A., 54, 62
Margolis, Julius, 95
Market data, use of, 77-96
Market demand and supply curves, 80-82
Market price, 77-78
Markets: imperfect, 83-85; nonexistent, 85-87; perfect, 77-83
Materials benefits, 2; estimating, 145-154
Materials damages: miscellaneous, 151-152; valuation of, 152-153
Materials, inventories of, 104, 148
McGlammery, G.G., 175, 188, 227
McIlvaine, Robert W., and Marilyn Ardell, 184
McKean, Roland N., 30, 95
Mechanical collectors, 215-218
Mendelsohn, Robert, 8, 59, 62
Mendelsohn, Robert, and Guy Orcutt, 105, 116-117, 120, 121, 129
Mera, Koichi, 36
Merrett, A.J., and Allen Sykes, 179
Metals dose-response functions, 146-149
Meteorological range, defined, 159
Michael Baker, Jr., Inc., 190
Michael, Robert T., and Gary S. Becker, 95
Michaelson, Irving, and Boris Tourin, 156-157, 158
Microepidemiologic studies, 112
Middleton, W.E.K., 159
Milk production, 140, 142, 202
Millecan, A.A., 139, 141

Miller, C. Arthur, 178
Mishan, E.J., 10, 62
Mist eliminator, 183
Model, descriptive/normative, 70
Model plants, 172
Morbidity studies, 117-118
Mortality studies, 113-117
Motor vehicle emissions, 171, 181; control systems, 194, 195
Mueller, W.J., and P.B. Stickney, 152
Multicollinearity, 108-109, 116, 130

National Academy of Sciences, 197
National income accounting, 36
National Institute for Occupational Safety and Health, 197, 199, 200, 210
National Research Council (NAS), 197, 198, 203, 210, 211, 212
Natural experiments, 112
Needleman, Lionel, 131
Nelson, Jon P., 165-166
Nelson, William C., John H. Knelson, and Victor Hasselblad, 122, 124
Net present value (NPV), 152-153
Net-present-value decision rules, 45-46
Net-social-benefit criterion, 57-60
Neveril, R.B., 171, 172, 176, 178, 189, 192-194, 195
Neveril, R.B., J.U. Price, and K.L. Engdahl, 195
New York City, 117
Nitrogen-dioxide control, 190
Nitrogen oxides, 111, 123, 124, 139, 145, 150, 203-205
North, D. Warner, and M.W. Merkhofer, 8, 124
Nylon, 150-151

Occupational exposures, 119-120, 121
Occupations, hazardous, 126-127
O'Connor, J.A., D.G. Parbery, and W. Strauss, 134
Odor damages, 155, 165, 202, 205-206
Olson, Craig A., 126-127, 131
Operating and maintenance (O & M) costs, 173, 184, 185, 189, 190, 191, 229-236
Opportunity-cost rate, defined, 39
Optimization over time, 37-38
Orange trees, 205
Order-of-magnitude estimates, 169-170; cost functions for, 181-189, 229-236
Ormrod, D.P., 134
Ornamental plants, 141-142
Oshima, R.J., 135
Ozone, 103, 111, 120, 123, 124, 134-135, 136, 138, 139, 143, 145, 149, 150, 152, 165, 206-207, 212

Page, Talbot, 10
Paint, 145, 154, 156, 167, 209-210, 214
Paint dose-response functions, 149-150
Palmini, Dennis, and Daniel Rossi, 194
Pareto optimality, 14
Park, William R., 179
Particulate-control devices, 190, 191, 215-223
Particulate Control Performance and Cost Model (Teknekron), 191
Particulate matter. *See* Total suspended particulate
Pasture, 138, 140, 142, 214
Payback-period decision rules, 48, 50-52
PEDCo Environmental, 191
Pesticides and herbicides, 210
Peters, Max S., and Klaus D. Timmerhaus, 179, 190
Philadelphia, 117, 156, 157
Phlips, Louis, 23
Pikulik, Arkadie, and Hector E. Diaz, 190
Plant damage, defined, 133
Plant injury, defined, 133
Plant sensitivity to insect damage, 140
Policy evaluation: and expected monetary value, 70-74
Pollutant exposure, estimating, 99-104
Pollution control equipment, 169, 215-227; cost functions, 184; lifetime, 176; purchases, 216
"Pollution-potential" estimates, 120
Pollution price differentials (real estate), 163-166
Polycyclic organic matter (POM), 211
Polyester, 150-151
Ponder, Thomas C., 190
Pooled data, 108
Population distributions, 103-104
Potatoes, 207
Potential-compensation criteria, 20-21, 26, 30-31; and economic-welfare criterion compared, 21-23
Potential economic welfare, 23-25
Preliminary estimate (control costs), 170
Present-value criterion, 41-45
Price deflator (control equipment), 178
Price indexes, 176-178
Private opportunity cost rate, 41, 53-54
Private rate of time preferences, 40-41, 53-54, 60
Producer's Price Index, 176, 188
Producer's surplus, 79-80, 174
Project lifetime, 176
Project scope, 175
Property-value studies, 163-166
Prospect theory, 69-70
Public-annoyance dose-response function, 155
Public good, 4
Purchase cost (pollution control), 170, 172

Radioactive matter, 105, 211-212
Rate of time preference, defined, 39
Rawls, John, 36
Rayon, 150
Real-estate values, 163-166
Redistribution, 24-25
Redundancy factor, 186
Regression analysis, multivariate, 107-108; sources of error, 108-110
Regulatory, monitoring, and enforcement costs, 174, 192
Replacement expenditures, 145, 152-153, 174
Respirable particulate matter, 120, 207-209
Respiratory diseases, 111-112, 118, 123, 200, 204, 206-207, 209, 211, 213
Retrofit factor, 175, 187
Return-on-investment ratio, 179
Ricci, Paolo F., and Ronald E. Wyzga, 113
Ridge regression techniques, 116
Ridker, Ronald G., 156, 165
Risk, 68, 126-128
Risk premium, 73-74
"Roll-back" model, 102-103
Rose-Ackerman, Susan, 10
Rowe, Robert D., Ralph C. d'Arge, and David S. Brookshire, 162-163
Rubber, 145, 152, 207
Rubinfeld, Daniel L., 165
Ruby, Michael G., 125
Rutledge, Gary L., 1
Rutledge, Gary L., and Betsy O'Connor, 1

Salmon, Richard L., 145
Salvin, Victor S., 154
Samuelson, Paul A., 10, 61
Sandmo, Agnar, and Jacques H. Dreze, 62
Scaling factors, 182-183
Scandinavia, 214
Schimmel, Herbert, and Thaddeus J. Murawski, 117
Schroeder, W.H., 206
Schwing, Richard C., 7, 128, 194
Schwing, Richard C., and Gary C. McDonald, 116, 120
Scitovsky, Tibor, 21
Scott, R., 195
Scrubbers, 182-184, 185, 222-224
Sen, Amartya K., 62
Senew, Michael, 192
Shawnee Lime-Limestone program, 190-191
Shy, Carl M., 112, 197
Sigmoid dose-response curve, 105-106

Index

Site adjustments, 175, 192
Skelly, John M., Sagar V. Krupa, and Boris I. Chevone, 140
Slovic, Paul, Sarah Lichtenstein, and Baruch Fischhoff, 128
Sludge handling and disposal, 190, 191
Smith, Barton A., 166
Smith, Robert S., 126
Smog, photochemical. *See* Ozone
Smoke plumes, 161
Smoking habits, 119-120, 121
Social opportunity cost rate, 40-41, 52-62
Social rate of time preference, 40-43, 52-62
Social value of investment, 54-57
Social welfare, defined, 13
Social-welfare criterion, 13-14, 17-19, 26
Social-welfare domination, 30-31
Soiling damages, 145, 155-157, 209; valuation of, 157-159
Soot, 211
Soybeans, 138
Spare-parts inventory, 192
Spence, James W., Fred H. Haynie, 154
Spengler, John D., 113
Spinach, 141
St. Louis, 165
St. Petersburg Paradox, 65
Stack, 226
"Startup" costs, 194
State of the world, 63
Steel, 146, 147-148, 214
Stern, Arthur C., 197, 227
Stevens, R.W., 176
Strategic Environmental Assessment System (SEAS), 183, 229-236
Stress-corrosion cracking, 147
Study estimates, 170, 181, 189-191
Sugar beets, 202
Sulfate, 113, 116, 118, 120-121, 123, 130, 160, 161, 212
Sulfation "candles" or "plates," 109, 121, 213
Sulfur-byproduct credit, 189
Sulfur oxides, 109-121, 136-152, 190, 191, 212-214
Sulfur removal rate, 187
Supplementary control system (SCS), 192
Supply curve, 79-80
Surveys, willingness to pay, 125-126, 161-163
Systems Application, Inc., 102

Tampa, Fla., 125, 161-162
Taxes, 84, 174, 180
Teknekron, 191
Tennessee Valley Authority, 191
Textiles dose-response functions, 145, 150-151
Thaler, Richard, and Sherwin Rosen, 126, 128, 131
Theil, Henri, 23
Thibodeau, Lawrence A., 116
Threshold of effects, 105-107, 121, 122-123
Tietenberg, Thomas H., 10
Time-series data, 108
Tingey, David T., and Richard A. Reinert, 138
Tires, vehicle, 152
Tomatoes, 135, 141
Torstrick, R.C., L.J. Henson, and S.U. Tomlinson, 191
Total cost (control equipment), 170
Total suspended particulate (TSP), 113-117, 120-121, 145, 156-158, 159-161, 165, 166, 167, 207-210
Tower absorbers, 183, 190, 224
Traffic flow improvements, 175, 194-195
Transformation curve, 38
Transportation-pollutants-control costs, estimating, 194-195
Trees, 140, 201-202, 207, 214
Turbulent-contact-absorber scrubber, 191
Turner, D. Bruce, 100, 101
Tversky, Amos, and Daniel Kahneman, 75

Uhl, Vincent, 180
Uncertain benefits and costs, evaluation of, 63-75
Uniformly distributed dollar (UDD), 32-33
U.S. Bureau of Economic Analysis (BEA), 178
U.S. Bureau of Labor Statistics, 178, 188
U.S. Council on Environmental Quality, 124
U.S. Department of Agriculture, 8
U.S. Department of Energy, 183
U.S. Department of the Treasury, 147, 176
U.S. Environmental Protection Agency (EPA), 8, 103, 104, 143, 150, 171, 179, 194, 195, 201, 203, 205, 207, 214; *Air Quality Criteria,* 197; Community Health and Environmental Surveillance System (CHESS), 118, 122, 213; "Cost of Clean Air" project, 183, 229-236; New Source Performance Standards (NSPS), 172, 189-190, 229
U.S. House of Representatives, 8, 118
U.S. Interagency Task Force, 123
U.S. Internal Revenue Service (IRS): asset depreciation guidelines, 147, 176
U.S. Water Resources Council, 103
University of Michigan, 118
Utility, defined, 13
Utility function, 67, 68-69

Utility measures, 26
Utility-possibility frontier, 16

Vegetation, animal, and ecosystem damages, valuation of, 141-142
Vegetation: injury to, 136-138, 201-202, 209; inventories, 104
Vegetation and ecosystems benefits, 2; estimating, 133-143
Vegetation dose-response functions, 133-140
Venturi scrubber, 183, 188, 190, 191, 194, 222-224
Vinyl chloride, 197
Viscusi, W. Kip, 126-127
Visibility, defined, 159
Visibility damages, 159-161, 210; valuation of, 161-163
Von Neumann, John, and Oskar Morgenstern, 66

Wage premiums, 126-128, 131
Waggoner, Alan P., and Ray E. Weiss, 160
Walsh, Phillip J., George G. Killough, and Paul S. Rohwer, 112
Washington, D.C., 165-166
Waste removal and treatment equipment, 226
Watson, William D., and John A. Jaksch, 157-158
Weaver, James B., and H. Carl Bauman, 170, 179, 195

Weisbrod, Burton A., 36
Weiss, Ray E., 160
Weitzman, Martin L., 10
"Welfare economics, new," 20, 27
Wellman-Lord process, 186, 187, 188, 189, 190, 191
Westman, Walter E., 142
Wheat, 124
Williams, Alan, 9
Williams, Roger, Jr., 195
Willig, Rogert D., 79, 95
Willig, Robert D., and Elizabeth E. Bailey, 30-31, 36
Willingness-to-pay, 125, 155, 161-163, 166
Wilson, Richard, 105
Wood siding, soiling damage, 156-157
Woods, Donald R., 179
Woods, Donald R., Susan J. Anderson, and Suzanne L. Norman, 189
Working capital, 192, 195
Wyzga, Ronald E., 117

Yen, Yen-Chen, 178
Yield dose-response functions (vegetation), 134-135
Yocom, John E., and Nicola Grappone, 145

Zeckhauser, Richard, and Donald Shepard, 128
Zielhuis, R.L., 199
Zinc, 146-147, 148-149

About the Authors

Robert Halvorsen is associate professor of economics at the University of Washington in Seattle. He received the B.B.A. from the University of Michigan and the M.B.A., M.P.A., and Ph.D. in economics from Harvard University. His principal areas of research are natural resources and environmental economics.

Michael Ruby is a consulting engineer in Seattle, Washington. He received the B.S. from the University of Oklahoma and the M.S., M.S.E., and Ph.D. in engineering from the University of Washington. His principal areas of work are air-pollution engineering and environmental economics.